THE HISTORY AND TOPOGRAPHY OF IRELAND

ADVISORY EDITOR: BETTY RADICE

GERALD OF WALES, one of the formidable Geraldines, a grandson of Gerald de Windsor and the Princess Nest, was born *c.* 1146 in Manorbier, Pembrokeshire. He died in obscurity in 1223, possibly in Lincoln. Three parts Norman and one part Welsh, he was one of the most dynamic and colourful churchmen of the twelfth century. His dream was that he might become Bishop of St David's, be consecrated without having to acknowledge the supremacy of Canterbury and then persuade the Pope to appoint him Archbishop of Wales. For this he fought with great courage and tenacity over the years, refusing four other bishoprics in Ireland and Wales, and preferring to remain Archdeacon of Brecon if he could not realize his grand design. He knew almost everyone worth knowing in his day, kings, popes, Welsh princes, prelates; he argued his case with most of them and he criticized them with some venom in his writings. He wrote seventeen books and planned a number of others, all in Latin. *The Journey through Wales* and *The Description of Wales* (Penguin Classics), obvious counterparts to each other, are among his more amiable works. Gerald was a member of one of the leading Norman families involved in the invasion of Ireland, which he first visited in 1183. His *History and Topography of Ireland* is an account of that country and its early history as seen by a Norman in A.D. 1185, and is an invaluable source of Irish history of the whole of the Middle Ages.

JOHN O'MEARA was born in Eyrecourt, County Galway, Ireland, in 1915 and was educated at University College, Dublin, and Oxford University. He was Professor of Latin at University College, Dublin, from 1948 to 1984, has been a Member of the Institute for Advanced Study, Princeton, N.J., and held a Fellowship from Harvard University at Dumbarton Oaks Byzantine and Mediaeval Humanities Research Center in Washington, D.C., from 1979 to 1984. He was Director of Studies on Johannes Scottus Eriugena at the Royal Irish Academy from 1984 to 1989, and has been a Research Associate in Classics at Trinity College, Dublin, since 1984. Among his published works are *The Young Augustine*, *Porphyry's Philosophy from Oracles in Augustine*, *Charter of Christendom: The Significance of the City of God* and *Eriugena*.

THE History and Topography OF IRELAND

GERALD OF WALES

Translated with an Introduction by
John J. O'Meara

PENGUIN BOOKS

PENGUIN BOOKS

Published by the Penguin Group
Penguin Books Ltd, 80 Strand, London WC2R 0RL, England
Penguin Putnam Inc., 375 Hudson Street, New York, New York 10014, USA
Penguin Books Australia Ltd, 250 Camberwell Road, Camberwell, Victoria 3124, Australia
Penguin Books Canada Ltd, 10 Alcorn Avenue, Toronto, Ontario, Canada M4V 3B2
Penguin Books India (P) Ltd, 11 Community Centre, Panchsheel Park, New Delhi – 110 017, India
Penguin Books (NZ) Ltd, Cnr Rosedale and Airborne Roads, Albany, Auckland, New Zealand
Penguin Books (South Africa) (Pty) Ltd, 24 Sturdee Avenue, Rosebank 2196, South Africa

Penguin Books Ltd, Registered Offices: 80 Strand, London WC2R 0RL, England

www.penguin.com

First published by the Dundalgan Press 1951
Revised edition published by the Dolmen Press 1982
Published in Penguin Books 1982
030

Printed in England by Clays Ltd, St Ives plc
Set in Filmset Ehrhardt

ISBN-13: 978–0–140–44423–0

www.greenpenguin.co.uk

CONTENTS

ILLUSTRATIONS

The twelfth-century map of Europe on page 21 and the line drawings in the text are copies of the coloured illustrations in the manuscript in the National Library of Ireland (MS 700).

The modern map on page 22 was drawn for this edition by Reg Piggott, by permission of the Ordnance Survey of Ireland.

FOREWORD

IN THIS NEW and revised edition of the translation published by the Dundalgan Press in 1951 of Giraldus Cambrensis' work, sometimes called *History*, sometimes *Description* and sometimes *Topography of Ireland*, I have thought it preferable to follow rather more closely my edition of the first recension of the Latin text in the *Proceedings of the Royal Irish Academy*, 52 C 4, 1949. I give, for example, the original list of contents (with its occasional discrepancies from the headings found in the body of the text), and the marginalia, which comment rather haphazardly on the text from time to time. I have also kept the proper names as they appear in the edition of the Latin text, except where to do so would seem rather artificial; but I have tried to help the reader, where he was likely to have difficulty, in the Notes. The choice of illustrations from MS 700 of the National Library has also been varied to harmonize with this edition.

Although my translation was more than favourably received, I have made a number of corrections and have had, of course, to add a few bibliographical items to give some direction to the reader of today. I gratefully record my indebtedness to the work of two colleagues in University College, Dublin: F. X. Martin, 'Gerald of Wales, Norman Reporter', *Studies*, 1969; and M. Richter, *Giraldus Cambrensis*, Aberystwyth, 1972, and 'Gerald of Wales', *Traditio*,

XXIX, 1973. The volume *Speculum Duorum* (of which M. Richter was general editor), Cardiff, 1974, has been very useful. I am particularly grateful to Dr Richter for suggested corrections incorporated in the present edition.

INTRODUCTION

GIRALDUS DE BARRI, called Cambrensis from Cambria (Wales), the country of his birth, was born at the castle of Manorbier, in Pembrokeshire, about A.D. 1146. His father was William de Barri, whose ancestors had taken their name from the little island of Barri, off the coast of Glamorganshire. His mother was Angharad, daughter of Nest who was the celebrated mistress of Henry I and was described as the 'Helen of Wales'. Through her marriage with Gerald de Windsor, castellan of Pembroke, Nest had become the mother of the first members of the illustrious and powerful Norman family of the fitzGeralds. Among her other children were David fitzGerald, later bishop of St David's, and Maurice fitzGerald, one of the principal leaders of the Norman invasion of Ireland. Here I describe no more of Giraldus' life and works than will provide a brief context for his *Topography of Ireland*.

Little is known of his early upbringing and education. Indeed J. S. Brewer in his introduction to the first volume of Giraldus' works (*Giraldi Cambrensis Opera*, Rolls Series, 1861) is of the opinion that Giraldus could have got very little learning in Wales, either from the Welsh or the Normans there. But he does tell us himself in his *De Rebus* that he received the equivalent of a high-school education under Master Haimon at the Benedictine abbey of Gloucester. From the beginning he was drawn to literature and an ecclesiastical career. He went to Paris in 1162, where he may have been affected by the literary and artistic movements of twelfth-century France, but where he certainly perfected his knowledge of the Latin poets and devoted himself to the study of law, philosophy, and theology. He returned home after some thirteen years in Paris and in 1175 became Archdeacon of Brecon and a very vigorous supporter of ecclesiastical discipline. In 1176 the see of St David's fell vacant. He was chosen by the chapter as bishop-elect among four names put forward, but was rejected by the king, who would not appoint a Welshman to a Welsh see. He thereupon returned to Paris, but in 1179 became administrator of the see, which he so much coveted, during the absence of the bishop. In 1183 he paid his first visit to Ireland where his family were the chief instruments in the conquest. He joined the entourage of Henry II in 1184, and was employed partly in diplomatic negotiations with the Welsh, and partly as tutor to Lord John, with whom he came to Ireland a second time in 1185. Under Richard I he was attached to the bishop of Ely who administered the realm during the absence of the king in the Holy Land. Subsequently he retired to Lincoln to resume his studies, emerging from time to time to refuse a number of bishoprics that were offered to him. However, he sought appointment to the see of St David's when once again it became vacant in 1198. Again, after four years of contention in which the Pope was involved, he was refused. Although the see was available once more in 1215 and may have been offered to him, he is said to have declined it. Giraldus paid a third visit to Ireland to see his cousin fitzHenry, Justiciar of Ireland, in 1199 and came again in 1204 when he stayed for some two years. He died in 1223.

The works of Giraldus have been edited by J. S. Brewer and J. F. Dimock in the Rolls Series (1861 on). The *Topography of Ireland*, edited by Dimock, is in Volume V, published in 1867. The materials for an up-to-date bibliography are to be found in the *Gemma Ecclesiastica*, translated by J. J. Hagen, Leiden, 1979, pp. 343 f. Among the works of Giraldus are the *Symbolum Electorum* (a collection of letters and poems), edited by M. Richter and translated by B. Dawson, Cardiff, 1974, some political tracts, the *De Rebus a se Gestis* (part of an autobiography), and a number of historical treatises, of which the *Topography of Ireland* was the first. Others were projected, but we now have only the *Conquest of Ireland*, of which a new critical edition by A. B. Scott and F. X. Martin was published in Dublin in 1978; and finally *The Journey through Wales* and *The Description of Wales*, translated by Lewis Thorpe, Penguin Books, 1978. Just as Giraldus gave 'The Prophetic History' as a sub-title to the *Conquest of Ireland* – in as much as he wrote it against the background of the prophecies of Merlin Silvester of Celidon (see n. 161 to the edition referred to above) – so in his address to Henry II at the beginning of the *Topography of Ireland* he refers to himself as Silvester, a 'prophet' in Henry's interest.

To the *Topography of Ireland*, as Brewer says (op. cit., I, xl), 'we are indebted for all that is known of the state of Ireland during the whole of the middle ages, a few barren Chronicles excepted'. Its great importance may also be seen from the remarks of a more interested writer, Geoffrey Keating, who says: 'Every one of the new Galls who writes on Ireland writes . . . in imitation of Cambrensis . . . because it is Cambrensis who is as the bull of the herd for them for writing the false history of Ireland, wherefore they had no choice of guide' (*History of Ireland*, Irish Texts Society, 1902, I, p. 153). We may be pardoned, then, if we say something on how the book was composed.

Giraldus, as we have seen, came to Ireland for the first time in 1183: 'he helped his uncle and his brother by his counsel, and diligently explored the site and nature of the island and primitive origin of its race' (*Conquest of Ireland*, Dimock, p. 351). His first stay, if not very short, was certainly not more than a year. On 25 April 1185 he landed in Ireland for the second time at Waterford. On this occasion he had been despatched by Henry II: 'The king sent his youngest son John to Ireland with a great force; and with him he sent Master Giraldus, because he had a great number of kinsmen in Ireland who had been among the first conquerors of that race, and also because he was an honest and prudent man' (ibid., pp. 380–81). After John's return to Wales and England in December of the same year, Giraldus was 'left with Bertrand of Verdun, the Seneschal of Ireland, to be his comrade and the witness of his deeds, and remained in the island to the following Easter, that he might pursue his studies more fully, not merely gathering materials but setting them in order' (*De Rebus*, I.65).

From this testimony, the internal evidence of the *Topography of Ireland*, and the known activities of the invaders, we can come to a fair idea of the extent of his travelling in, and knowledge of, Ireland. This is important in view of his confident and uncomplimentary assertions on many points. We have no proof that he went outside the neighbourhoods of Cork and Waterford on the occasion of his first visit to the country. On the second occasion, when he came over with John, he travelled from Waterford to Dublin. He can scarcely have gone by an inland route or he would hardly have written: 'On the whole the land is low-lying on all sides and along the coast; but further inland it rises up very high to many hills and even high mountains' (see p. 34). He may have journeyed by the coast, or even taken ship. He does seem, however, to have some knowledge of Arklow and Wicklow. We know, too, that he passed through Meath in 1185 (see p. 72), and was in Kildare (see Dimock, p. 124). He may have seen the Shannon near Athlone, with Lough Ree; perhaps also Lough Derg. That is all that we can reasonably conclude. The rest of the country was probably unvisited by him, as most of it was, up to that time, by his fellow Normans.

The result of his literary work during 1185 and the following two or three years was his first account of Ireland and its early history, his *History* or *Topography of Ireland*. The less accurate term *Topography* was used by himself when referring to it (cf. Dimock, pp. 7, 207) and this title persisted. It was read by, or to, Archbishop Baldwin during the seven or eight weeks of his preaching of a crusade in Wales, beginning in the middle of March, 1188. It may already have been read publicly at Oxford in the autumn of the previous year. In any case it was so read at Oxford in or about 1188.

Archbishop Baldwin thought that the *Topography of Ireland* was very good. Giraldus tells us that he never tired at its reading. He read it thoroughly. The bishop was particularly attracted by the elegance of the style, and the aptness of the allegorical moralizations. He asked Giraldus if he had made much use of earlier commentators or hagiographers. Giraldus answered that he had not; he had relied solely on the inspirations of God's grace. The good archbishop was duly impressed. The author, proud of his success, unfortunately set out to improve on it by filling out his chapters with still more symbolisms, moralizations, theological excursions, quotations from early writers (with comments on them), legendary accounts of other countries, indiscriminate erudition of all kinds, and well-pointed laudations of Henry II and his sons. As Dimock says (p. xiv), 'they have about as much to do with Ireland or its people as with the moon and the man in it'. During twenty to thirty years, from the period, that is, when he retired temporarily to Lincoln in 1194 to his death in 1223, he continued to add to his original text, going through at least four recensions, until in the end the final version was more than twice as long as that read to Baldwin or at Oxford.

The first recension is now, for the first time, given here in English. Of its presentation at Oxford Giraldus himself tells us:

> When in process of time the work was finished and corrected, and not wishing to place the candle which he had lit under a bushel but to lift it aloft on a candlestick that it might shine forth, he determined to read it before a great audience at Oxford, where, of all places in England, the clergy were most strong and preeminent in learning. And since his book was divided into three

parts (*distinctiones*), he gave three consecutive days to the reading, a part being read each day. On the first he hospitably entertained the poor of the whole town whom he had gathered together for that purpose; on the morrow he entertained all the doctors of the divers faculties, and those of their scholars who were best known and best spoken of; and on the third day he entertained the remainder of the scholars together with the knights of the town and a number of citizens. It was a magnificent and costly achievement, since thereby the ancient and authentic times of the poets were in some manner revived, nor has the present age seen nor does any past age bear record of the like (*De Rebus*, p. 73).

It will be seen that the merits of the original work are considerable. The whole story gains immensely in vigour and interest by the omission of extraneous matter which to a modern reader can only be tedious.

The text here translated, however, is not arrived at by any process of cutting down on the final account. We are fortunate in having three manuscripts of the original work as penned by Giraldus. The manuscripts are called M, P, and H. M (Mm. 5.30, University Library, Cambridge) is the earliest and most valuable. It is a twelfth-century manuscript but can scarcely have been written by Giraldus himself, for it reveals itself as a copy in showing obvious blunders and omissions. It is the basis of the recension (see O'Meara, *Proceedings of the Royal Irish Academy*, 52 C 4, 1949) from which this translation is made. P (MS 181, Peterhouse, Cambridge) is a fifteenth-century copy of a copy of the original manuscript independent from M. H (Harleian 3724, British Museum) is a thirteenth-century manuscript, but is less valuable than P. All the materials for a study of the question are available in Dimock's edition of *Giraldi Cambrensis Opera* (Rolls Series, 1867), V.

Of one other manuscript I should like to make particular mention, although it is of little interest as a text. It was formerly Phillipps MS 6914, but is now MS 700 of the National Library of Ireland, by which it was acquired in 1946. From a note on this manuscript we learn that it was given by Walter Mybbe to the vicars of Hereford Cathedral in 1438. It is just possible, as A. Gwynn tentatively suggested to me, that it is the very copy mentioned by Giraldus himself in the *De Rebus*, p. 409. The map and some of the very spirited illustrations from this manuscript are reproduced here.

On the question of the sources of his information for the *Topography of Ireland* Giraldus boasts that he used no written sources whatever for the first two parts, but he admits that for some of the third part he is indebted for a little information to the unhelpful chronicles of the Irish. The chronicles to which he refers certainly include the oldest extant version of the *Lebor Gabála*. But see nn. 5, 15, and 42.

The *Topography of Ireland* has been very severely and rightly criticized not only by Irishmen but by others as well. Giraldus himself in his *Expugnatio* considered it necessary to answer contemporaries who refused to accept all he had written. A full list of the replies to the book will be found in Dimock (op. cit., lxxv–lxxxii). The most important are those of the Jesuit, Stephen White, between 1603 and 1607: *Apologia pro Hibernia* (printed by Matthew Kelly, Dublin, 1849); and Dr John Lynch, 1662: *Cambrensis Eversus* (edited by Matthew Kelly, Dublin, 1848–52).

The reader will be able to judge for himself the amount of credit to be placed in Giraldus' statements, and the motives by which he was actuated. He will see the single-minded vanity of an ambitious flatterer, the haughty contempt of one who came with his family to reform and invade, and the apparent credulity which must have delighted the hearts of the Irish. It is usual to use hard words of Giraldus. And yet, when he has been taken to task for his excesses, it is well to remember that without his *Topography of Ireland* and *Conquest of Ireland* our knowledge of the Ireland of the twelfth century would be much the poorer. Moreover, he has the gift of story-telling and a lively, if too rhetorical, style.

This translation is meant, not for the scholar, but for the ordinary reader. Hence the notes are few and do not include comprehensive reference to Biblical or other sources. The aim has been to translate accurately and to preserve the *flavour* and directness of the original work as much as possible.

Finally, I should like to express my thanks to Rev. Professor Gwynn, S.J., Rev. Professor Ryan S.J., and Dr Bieler, for much help and advice.

GUIDE TO MAP OF EUROPE *C.* 1200 A.D. See pp. 16–17

The guide follows the map, generally, from the bottom upwards and from left to right within the sections divided by water. The top left section, however, is taken last. ¯ indicates an abbreviation.

Ancient name	*Modern name*
ẎSLANDIA	ICELAND
HẎBERNIA	IRELAND
Sinnenus	Shannon
Auenliff̄	Liffey
Dubliñ	Dublin
Slana	Slaney
Weiseford̄	Wexford
Suir̄	Suir
Watr̄fordia	Waterford
Limericū	Limerick
Orcades	Orkney Islands
SCOTIA	SCOTLAND
Mare Scotic̄	Firths of Clyde and Forth
WALLIA	WALES
Sabrina	Severn
BRITĀNIA	BRITAIN
Eborac̄	York
Humbra	Humber
Lincoln	Lincoln
Lundoñ	London
Tamisia	Thames
Winconia	Winchester
NORWAGIA	NORWAY
DACIA*	TRANSYLVANIA
SAXONIA	SAXONY
FRISIA	FRISIA
Colonia	Cologne
Renus	Rhine

Ancient name	*Modern name*
ALEMĀNIA	GERMANY
Remis	Reims
Atrabatum	Arras
FLANDRIA	FLANDERS
Hora Gallici-litoris	Gallic coastline
Secana	Seine
B̄GVND̄	BURGUNDY
Lugdunum	Lyons
FRANCIA	FRANCE
Parisii	Paris
Ligeris	Loire
NORMĀNIA	NORMANDY
Rotomag̃	Rouen
ARMORICA	ARMORICA
BRITĀNIA	BRITTANY
PICTAVIA	POITOU
GASCOÑ	GASCONY
HẎSPAÑ	SPAIN
PROVINCIA	PROVENCE
ANDEGAVIA	ANJOU
Turoñ	Tours
Andegavis	Angers
ALPES	ALPS
Lac̄ Losañ	Lake of Geneva
Rodanus	Rhone
Vienna	Vienne
Arausis	Orange
Arelas	Arles

Ancient name	*Modern name*	*Ancient name*	*Modern name*
ITALIA	ITALY	SICILIA	SICILY
Mediolanum	Milan	Siñ Liburnič	Gulf of Kvarner
Padus	Po	Dalmatia	Dalmatia
Placentia	Piacenza	Macedonia	Macedonia
Pavia	Pavia	Tracia	Thrace
Svsa	Susa	Constātinopol̄	Constantinople
Mons Apenniñ	The Apennines	Dardania	Part of Serbia
TVSCIA	TUSCANY	Mesia	Serbia and N.W. Bulgaria
ROMA	ROME		
Tiberis	Tiber	Danubius	Danube
Apulia	Apulia	Bavaria	Bavaria
Boneuentum	Benevento	Boemia	Bohemia
Campania	Campania	Vngaria	Hungary
Lvcania	Lucania	Sicābria*	N.W. of Cologne
Baris	Bari	Gotia	Area near Sevastopol
Calab̄a	Calabria		
Regiū	Reggio	Theodosia	Feodosiya

*Dacia and Sicambria should possibly be interchanged on the map.

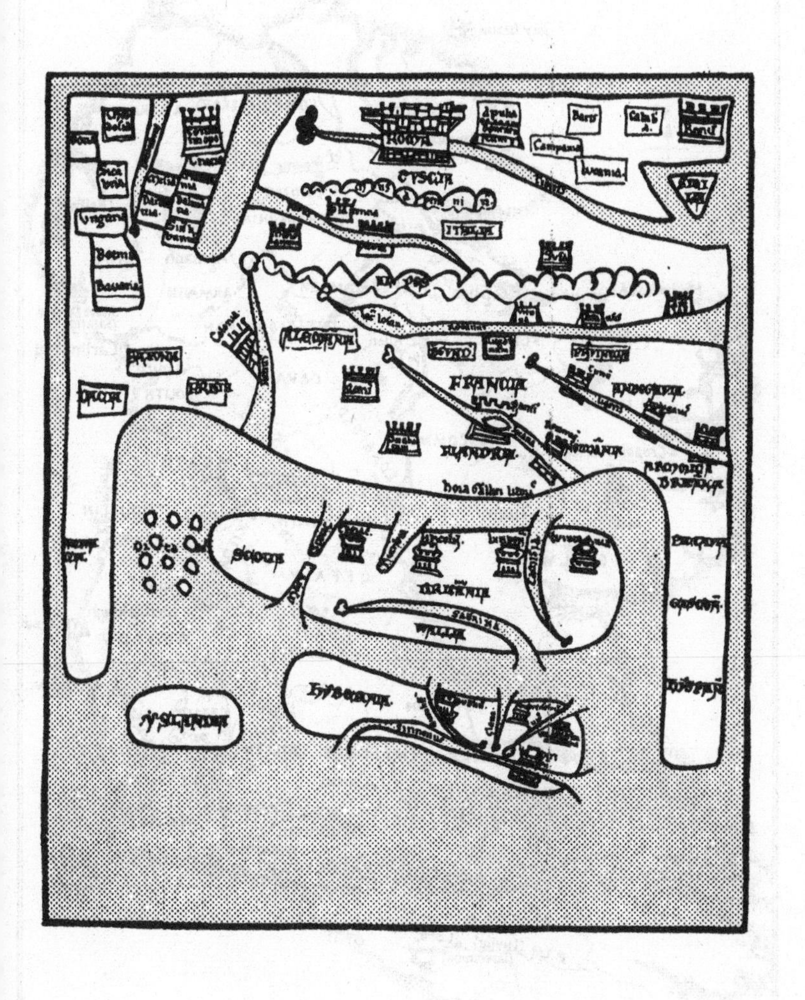
Ungaria
Boemia
Bavaria
Campania
Roma
Tuscia
Italia
Francia
Flandria
Wallia

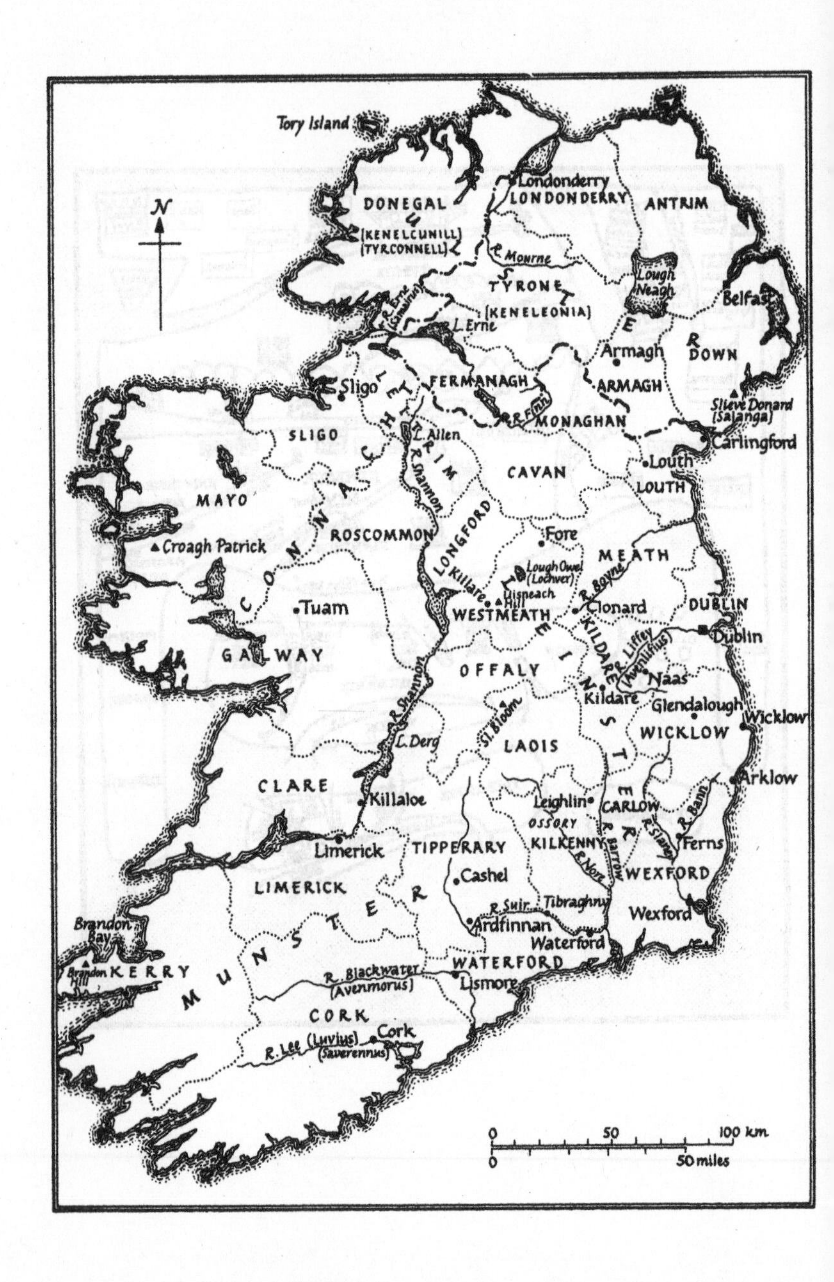

Tory Island
N
Londonderry
LONDONDERRY
ANTRIM
DONEGAL
(KENELCUNILL)
(TYRCONNELL)
R. Mourne
Lough Neagh
Belfast
TYRONE
(KENELEONIA)
R. Erne
L. Erne
ULSTER
Armagh
DOWN
Sligo
FERMANAGH
ARMAGH
Slieve Donard (Salanga)
R. Finn
MONAGHAN
SLIGO
L. Allen
LEITRIM
Carlingford
R. Shannon
CAVAN
Louth
LOUTH
MAYO
CONNACHT
ROSCOMMON
LONGFORD
Fore
Croagh Patrick
MEATH
Lough Owel (Lochver)
Killare
Uisneach Hill
R. Boyne
Tuam
WESTMEATH
Clonard
DUBLIN
KILDARE
Dublin
R. Liffey (Avenlifius)
GALWAY
OFFALY
Naas
LEINSTER
Kildare
Glendalough
Wicklow
Sl. Bloom
WICKLOW
L. Derg
LAOIS
R. Shannon
Arklow
CLARE
Killaloe
Leighlin
CARLOW
R. Bann
OSSORY
R. Slaney
Limerick
TIPPERARY
KILKENNY
R. Barrow
Ferns
R. Nore
Cashel
WEXFORD
LIMERICK
R. Suir
Tibraghny
MUNSTER
Ardfinnan
Wexford
Brandon Bay
Waterford
Brandon Hill
KERRY
WATERFORD
R. Blackwater (Avenmorus)
Lismore
CORK
Cork
R. Lee (Luvius) (Saverennus)
0
50
100 km
0
50 miles

¶ Giraldus Cambrensis on the Topography of Ireland

The book falls into three parts

THE FIRST PART treats of the Position of Ireland. The distance between Ireland and Britain. What land it has to the south and how far distant. What land to the north, what land to the east, what to the west. What proportion Ireland bears to Britain in size. The longitude and latitude of Ireland. The character of the country and its unevenness. The fertility of the tillage land and the smallness of the grains of wheat scarcely to be separated from the chaff by the winnowing fan. Exposure to wind and rain. The prevailing north-west wind that bends the trees. The opinions of Bede and Solinus on vineyards and bees. The nine principal rivers and many others that have recently emerged. The lakes and their islands. The sea, river and lake fish and those that are missing. Fish that are new and not found elsewhere. Birds and those that are missing. The hawk, falcon and sparrow-hawk and their nature. The eagle and its nature. The crane and its nature. Barnacles that are born of the fir-tree and their nature. Birds that belong to two species and their natures. Kingfishers and their natures. Swans and storks and their natures. The many kinds of crows here and their natures. Wild animals and their natures and those that are missing. The badger and its nature. The beaver and its nature. Reptiles and those that are missing and how all poisonous reptiles are not found here. Poisonous reptiles when brought here immediately die and poison loses its force. The dust of this land which kills poisonous reptiles. The boot-thongs of this country used as antidotes against poisons. A frog discovered in Ireland. The many good points of the island and the natural qualities of the country. The advantages of the West are to be preferred to those of the East. All the elements in the East are pestiferous. The venom of poisons there and the harshness of the climate. The incomparable mildness of our climate. Certain deficiencies here that are in fact praiseworthy. The well of poisons is in the East.

THE SECOND PART treats of the Wonders and Miracles of Ireland. The surging tides of the Irish Sea and its varying ebb and flow. The contrary motions of the tides in Ireland and Britain. The moon affects the waters as well as natural humours. Two islands in one of which no one dies; into the other no animal of the female sex can go. An island one part of which is frequented by good and the other by evil spirits. An island where human corpses exposed in the open do not putrefy. The wonderful natures of wells. Two wonderful wells, one in Brittany and the other in Sicily. A big lake that had a marvellous origin. A fish with three gold teeth. The northern islands almost all of which are held by the Norwegians. The island that was at first unstable and then was made stable by means of fire. Iceland whose people speak few words but true and never swear on oath. A whirlpool of the sea that swallows ships. The Isle of Man which on account of its admission of poisonous reptiles is judged to belong to Britain. That a long time after the Flood and then not suddenly but gradually and as it were through inundation islands came to be. Thule an island of the West very well known in the East but entirely unknown in the West. The Giants' Dance which was transferred from Ireland to Britain. The wonderful happenings of our own time; and first about a wolf that talked with a priest. A woman with a beard and a mane on her back. A man that was half an ox and an ox that was half a man. A cow that was partly a stag. A goat that had intercourse with a woman. A lion that loved a woman. How the cocks in Ireland crow differently from those in other countries. Wolves that whelp in December. Ravens and owls that have their young about Christmas time. About miracles; and first about the fruit and ravens and blackbird of Saint Kevin. The teal of Saint Colman which are almost tame but fly away from hurt. The stone which every day miraculously contains wine. The fleas that were banished by Saint Nannan. The rats that were expelled from Ferneginan by Saint Yvor. The fugitive bell. Various miracles in Kildare; and first about the fire that never goes out and whose ashes do not increase. How Brigid keeps the fire on her own night. The hedge around the fire that no male may cross. The falcon in Kildare that was tamed and domesticated. A book miraculously written. The composition of the book. The cross in Dublin that speaks and gives testimony to the truth. How the same cross became immovable. How a penny offered to the cross twice jumped back and on the third occasion after confession remained; and of the iron greaves that were miraculously restored. The fanatic at Ferns who foretold the future from the past. How an archer who crossed the hedge of Brigid went mad and how another lost his leg. How seed wheat on being cursed by the bishop of Cork did not grow, and in the following year was miraculously interchanged with rye. How Philip of Worcester was struck with sickness at Armagh, and how Hugh Tyrrell was scourged through divine intervention. The mill that will not grind on Sundays or anything stolen or plundered. The mill that women do not enter. How two horses on eating oats stolen from the same mill immediately died. That the saints of this country seem to be of a vindictive cast of mind.

THE THIRD PART treats of the Inhabitants of the Country. The first arrival that namely of Cesara the grand-daughter of Noah before the Flood. The second arrival that namely of Bartholanus (Parthalón) three hundred years after the Flood. The third arrival that namely of Nemedus from Scythia with his four sons. The fourth arrival that namely of the five brothers and sons of Dela who first divided Ireland into five portions. The first king of Ireland, Slanius. The fifth arrival that namely of the four sons of king Milesius from Spain, and how Herimon and Heberus divided the kingdom between them. The discord between the two brothers and how Heberus having been killed Herimon was the first king of the Irish people. Gurguintius king of the Britons who sent the Basclenses to Ireland and gave it to them to settle in. The nature, customs and characteristics of the people. The advantages and effects of music. The incomparable skill of the people in musical instruments. How many kings reigned from Herimon to the coming of Patrick and how the island was converted to the Faith by him. How there were no archbishops in Ireland before the arrival of John Papiro who established four *pallia* in the island. How the three bodies of Patrick, Columba and Brigid were in our own times found in Ulster in the city of Down and were translated. The Irish are ignorant of the rudiments of the Faith. Their vices and treacheries. How they always carry an axe as if it were a staff in their hand. A proof of their wickedness and a new way of making a treaty. The Irish are by nature's gift handsome but shameful in their practices and culture. A new and outlandish way of confirming kingship and dominion. Many in the island have never been baptized, and have not yet heard of the teaching of the Faith. The Irish clergy, in many points praiseworthy. The prelates should be reproved for their neglect of their pastoral office. How the clergy differ from monks and are to be placed above them. When monks are chosen as clerics they should fulfil the functions of a cleric. Prelates taken into the clergy from monasteries have some obligations as monks and some as clerics. A sly reply of the archbishop of Cashel. About bells, croziers and other such relics of the saints regarded by the people of Wales as well as by the people of Ireland with great reverence, and about a priest stricken with a double sickness. The great numbers among this people that are maimed in body. How many kings reigned from the time of Patrick to the coming of Turgesius. How during the reign of Fedlimidius the Norwegians under the leadership of Turgesius conquered Ireland. How the English say that it was Gurmundus, the Irish Turgesius that subjugated Ireland. How Gurmundus having been killed in Gaul Turgesius died in Ireland, having been deceived by what appeared to be girls. Whence Gurmundus came into Ireland or Britain. The wily question of the king of Meath. About the Norwegians who ruled for some thirty years and how they were driven out of Ireland. The coming of the Ostmen. How many kings reigned in Ireland from the death of Turgesius to Rothericus the last king of Ireland. How many kings there were from the first, Herimon, until the last, Rothericus. How the Irish people from the time of its first coming until the time of Gurmundus and Turgesius, and from their deaths until Henry the Second, the king of the English, remained unconquered. The victories of Henry the Second, king of the English. A brief summary of the same king's various titles and triumphs.

Here Begins the Account
OF
The Wonders
OF
IRELAND

by Giraldus Cambrensis

His SILVESTER
to HENRY THE SECOND
Illustrious King of the English

IT PLEASED YOUR EXCELLENCY, *invincible king of the English, duke of Normandy, count of Anjou and Aquitaine to send me from your court, with your beloved son John, to Ireland. And there, when I had seen many things not found in other countries and entirely unknown, and at the same time worthy of some wonder because of their novelty, I began to examine everything carefully: what was the position of the country, what was its nature, what was the origin of the race, what were its customs; how often, and by whom, and how, it was conquered and subjugated; what new things, and what secret things not in accordance with her usual course had nature hidden away in the farthest western lands? For beyond those limits there is no land, nor is there any habitation either of men or beasts – but beyond the whole horizon only the ocean flows and is borne on in boundless space through its unsearchable and hidden ways.*

Just as the countries of the East are remarkable and distinguished for certain prodigies peculiar and native to themselves, so the boundaries of the West also are made remarkable by their own wonders of nature. For sometimes tired, as it were, of the true and the serious, she draws aside and goes away, and in these remote parts indulges herself in these secret and distant freaks.

I have, therefore, collected everything, and have chosen out some of them. Those which I have thought worthy of being remembered I have, I hope usefully, put together and propose them for your attention – which scarcely any part of history escapes. I could, as others have done, have sent your Highness some small pieces of gold, falcons, or hawks with which the island abounds. But since I thought that a high-minded prince would place little value on things that easily come to be – and just as easily perish – I decided to send to your Highness those things rather which cannot be lost. By them I shall, through you, instruct posterity. For no age can destroy them.

¶ Here begins the first part of the History of Ireland

1 ¶ *The Position of Ireland*

IRELAND, THE LARGEST ISLAND beyond Britain, is situated in the western ocean about one short day's sailing from Wales, but between Ulster and Galloway in Scotland the sea narrows to half that distance. Nevertheless from either side the promontories of the other can be fairly well seen and distinguished on a fine day. The view from this side is rather clear; that from the other, over such a distance, is more vague. This farthest island of the west has Spain parallel to it on the south at a distance of three ordinary days' sailing, Greater Britain on the east, and only the ocean on the west; but on the northern side, at a distance of three days' sailing, lies Iceland, the largest of the islands of the north. Ireland, then, lies parallel to Britain in such a way that if you sail to the west from any British port you will meet Ireland at some point.

The distance between Ireland and Britain.

What land it has at a distance to the south.

Nevertheless Britain is twice the size of Ireland. For Britain from south to north is eight hundred miles long, and about two hundred miles broad; while Ireland in the same way stretches in length from the Brendanican mountains to the island of Columba that is called Torach – that is, a distance of eight days at forty miles a day; and in breadth stretches from Dublin to

Britain is double the size of Ireland.

the hills of Patrick and the sea beyond Connacht – that is, a distance of four days. Ireland is, however, in proportion to its size, more round. Britain is seen to be more oblong and narrow.

The character of the country and its unevenness.

Ireland is a country of uneven surface and rather mountainous. The soil is soft and watery, and there are many woods and marshes. Even at the tops of high and steep mountains you will find pools and swamps. Still there are, here and there, some fine plains, but in comparison with the woods they are indeed small. On the whole the land is low-lying on all sides and along the coast; but further inland it rises up very high to many hills and even high mountains. It is sandy rather than rocky, not only on its circumference, but also in the very interior.

2 ¶ *The fertility of the tillage-land and smallness of the grains of wheat*

THE LAND IS FRUITFUL and rich in its fertile soil and plentiful harvests. Crops abound in the fields, flocks on the mountains, and wild animals in the woods. The island is, however, richer in pastures than in crops, and in grass than in grain. The crops give great promise in the blade, even more in the straw, but less in the ear. For here the grains of wheat are shrivelled and small, and can scarcely be separated from the chaff by any winnowing fan. The plains are well clothed with grass, and the haggards are bursting with straw. Only the granaries are without their wealth. What is born and comes forth in the spring and is nourished in the summer and advanced, can scarcely be reaped in the harvest because of unceasing rain.

Exposure to rain.

For this country more than any other suffers from storms of wind and rain.

A north-west wind, along with the west wind to its south,

prevails here, and is more frequent and violent than any other. It bends in the opposite direction almost all the trees in the west that are placed in an elevated position, or uproots them.

The trees in the west bent by the north-west wind.

The island is rich in pastures and meadows, honey and milk, and wine, but not vineyards. Bede,[1] however, among his other praises of the island says that it is not altogether without vineyards. On the other hand Solinus[2] says that it has no bees. But if I may be pardoned by both, it would have been more true if each of them had said the opposite: it has no vineyards, and it is not altogether without bees. For the island has not, and never had, vines and their cultivators. Imported wines, however, conveyed in the ordinary commercial way, are so abundant that you would scarcely notice that the vine was neither cultivated nor gave its fruit there. Poitou out of its own superabundance sends plenty of wine, and Ireland is pleased to send in return the hides of animals and the skins of flocks and wild beasts. Ireland, as other countries, has bees that produce honey; but the swarms would be much more plentiful if they were not frightened off by the yew-trees[3] that are poisonous and bitter, and with which the island woods are flourishing. It is possible, of course, that in Bede's time there were, perhaps, some vineyards in the island; and some people say that it was Saint Dominic of Ossory[4] who brought bees into Ireland – and that was long after the time of Solinus.

The opinions of Bede and Solinus on vineyards and bees.

Neither would it be strange if these authors sometimes strayed from the path of truth, since they knew nothing by the evidence of their eyes, and what knowledge they possessed came to them through one who was reporting and was far away. For it is only when he who reports a thing is also one that witnessed it that anything is established on the sound basis of truth.

3 ¶ *The nine principal rivers and many others that have recently emerged*[5]

THE ISLAND IS DIVIDED and watered by nine principal and magnificent rivers, which have been famous from the earliest time of the earliest inhabitant of the land after the Flood, that is Bartholanus. Their names are: Avenlifius through Dublin; Banna through Ulster; Moadus through Connacht; Slicheius and Samairus through Kenelcunill; Modarnus and Finnus through Keneleonia; and Saverennus and Luvius through Cork.

There are many other rivers flowing through Ireland which are, so to speak, new, and with regard to the ones mentioned, only recently emerged. They are not, however, smaller than the former, and only on the point of antiquity are they inferior. Some of them spring from the bowels of the earth and from the veins of wells. Others arise suddenly from lakes. All of them are fine rivers and divide and separate the island into its sections with their long courses. I have thought it useful to say something about some of these. From underneath the foot of the mountain Bladma (Slieve Bloom) three large rivers spring. They are called the three sisters, because they have shared between them the names of three sisters, namely, Berva which flows through Leighlin; Eoyrus through Ossory; and Suirus through Ardfinan and Tibraccia to join the sea at Waterford. The Slana flows through Wexford; the Boandus through Meath; the Avenmorus through Lismore; and the Sinenus through Limerick.

The Shannon (Sinenus) rightly holds the chief place among all the rivers of Ireland whether old or new, both on account of the magnificence of its size, its long meanderings, and its abundance of fish. It rises in a certain large and beautiful lake that divides Connacht from Munster, and sends two arms, so to speak, in opposite directions of the world. One arm goes south, flows beside Killaloe, takes in Limerick, and, separating the two Munsters from one another for a distance of one hundred miles and more, pours itself into the Brendanican sea. The other arm, of equal importance, separates Meath and the farther parts of Ulster from Connacht, and after many wanderings eventually joins the northern sea. It, therefore, separates the fourth and western part of the island from the other three. It runs between them and marks Connacht off from sea to sea.

Of old the country was divided into five almost equal parts, namely: two Munsters, north and south, Leinster, Ulster, and Connacht. The prophecy of Merlin predicted that they would all be reduced to one. But we shall say more about this when we come to treat of it.

4 ¶ *The lakes and their islands*

THIS COUNTRY, above all others that we have seen, is well supplied also with beautiful lakes, full of fish and very large. They are a kind of speciality here. They contain islands rising to some height and very beautiful. The lords of the land usually appropriate them as places of safety and refuge, as well as of habitation. They are inaccessible except to boats.

5 ¶ *The sea, river and lake fish*[6] *and those that are missing*

THE SEA COASTS on all sides abound sufficiently with sea-fish. The rivers, however, and the lakes are rich in fish peculiar to themselves, and especially in fish of three kinds, namely, salmon, trout, and mud-eels. The Shannon abounds in sea lampreys. They serve as luxuries for the rich. But some fine fish,

found in other regions, and some magnificent freshwater fish are wanting. I mean pike, perch, roach, gardon, and gudgeon. Minnow, loach, bullheads, *verones*, and nearly all that do not have their seminal origin in tidal rivers are absent also.

6 ¶ *Fish that are new and not found elsewhere*

ON THE OTHER HAND the lakes of this country contain three kinds of fish that are not found anywhere else. There is one kind longer and more round than the trout. It has firm white flesh, and is pleasing to the taste. It is very like the tymal, except that the head is larger. There is another kind very like the sea-herring in shape, size, colour, and taste. A third kind is in every detail like the trout, except that it has no spots. These three kinds appear only in the summer and never in the winter. In Meath, near Fore, there are three lakes near one another, of which each has one kind of these fish. Neither of the other two kinds ever approaches it, even though the lakes communicate by a river that joins them. Moreover, if a fish of one lake is carried to the place and lake of another, it either dies or returns to its first home.

7 ¶ *Birds and those that are missing*

SOME BIRDS that live on the water and build their nests in high places are found here as elsewhere. But other types of birds have been altogether wanting even from the earliest times.

8 ¶ *The hawk, falcon and sparrow-hawk and their nature*

THIS COUNTRY above any other produces hawks, falcons and sparrow-hawks abundantly. These birds have been provided by nature with courageous hearts, curved and sharp beaks, and feet armed with talons, most suitable for catching their prey – and all to afford amusement to the nobles. There is one remarkable thing about these birds, and that is, that no more of them build nests now than there did many generations ago. And although their offspring increases every year, nevertheless the number of nest-builders does not increase; but if one pair of birds is destroyed, another takes its place. The nests do not increase in number for a variety of reasons; no reason, however, tends to their increase.

Cassiodorus tells us:[7] birds of this kind 'that live by prey, throw their weak young offspring out of their nests, lest they should accustom themselves to soft living, and beat them with their wings, and compel their young chicks to fly, so that they may turn out to be such that their parents can depend upon them'. When, as time goes on, their wings have got some strength, they are trained with the help of nature to prey, and are brought from their homes, and compelled by their faithless parents to remain away forever.

We notice one thing in particular about sparrow-hawks, and that is, that some of them are marked with white spots, some with red, and others with spots that are half white, half red. Some people, accordingly, think that this variety of colour arises from the kinds of trees in the place where they are born. But since on the same tree and in the same nest you will very often see young birds showing this variety of colouring, the view that this diversity arises from the parents in a natural way is more likely.

Here too, as in almost all other kinds of animals, nature makes the males stronger; but in these, and in such others as live by prey, and have to pursue their game, and who especially have need of strength and violence, the males have proportionately less superiority in as much as the females are stronger and more virile.

9 ¶ *The eagle and its nature*

YOU WILL SEE as many eagles here as you will kites elsewhere. This bird can look straight at the very rays of the sun with all its brightness, and so high does it soar in its flight that its wings are scorched by the burning fires of the sun. It lives so long that it seems to contend with immortality itself in the renewal of its youth.

So also can contemplative men gaze fully with their minds, and without any turning away, upon the very nature itself of the divine majesty and the true sun of justice; and when they put their hands to the plough of the heavenly paradise, do not look back. And those too who, trying to unravel in the sacred scriptures what is hard and secret in the celestial mysteries, reach beyond the limits allowed and which should not, and cannot be passed, fall back upon themselves, and remain below as if the wings of presumptuous intelligence, on which they were borne, had been burned. Since therefore the heaven of heavens is the Lord's and what is left over will be burned by fire, we ought to be mindful, and not ungrateful, that we have been admitted to a part of knowledge and not to the fullness of intelligence and understanding. So do holy men, having truly renewed the innocence of their youth and having already put off the old man and put on the new, arrive happily at the reward of eternal life.

10 ¶ *The crane*[8] *and its nature*

CRANES ARE SO NUMEROUS that in one flock alone you will often see a hundred or about that number. These birds, by a natural instinct, take their turns by night in watching the common safety, standing on one leg only, while in the other featherless claw they hold a stone. They do this so that if they should go to sleep, they will be wakened again immediately by the fall of the stone and continue their watch.

We should follow the example of this bird in watching and being on our guard, because we are entirely ignorant of the hour at which the thief will come. Some sacred duty should occupy our minds which, like the stone, will shake off torpor and allow us to think of nothing else but itself. If then it should by chance slip from our mind sometimes, the mind, not being used to being without it, will, as it were, wakened from its sleep, take it up again.

This bird has such a warm and fiery liver that, if it should eat iron, it will not let it through undigested.

So too bowels aflame with the fire of charity will tame and soften hearts that were once as hard and unyielding as iron, and compel and soften them to the union of brotherly peace.

Wild peacocks abound here in the woods. Wild hens which are commonly called *grutae* are few here and rather small, and very like partridges both in size and colour. Quail also is plentiful here. Hoarse and noisy *ratulae* are innumerable.

11 ¶ *Barnacles that are born of the fir-tree and their nature*

THERE ARE MANY BIRDS HERE that are called barnacles, which nature, acting against her own laws, produces in a wonderful way. They are like marsh geese, but smaller. At first they appear as excrescences on fir-logs carried down upon the waters. Then they hang by their beaks from what seems like sea-weed clinging to the log, while their bodies, to allow for their more unimpeded development, are enclosed in shells. And so in the course of time, having put on a stout covering of feathers, they either slip into the water, or take themselves in flight to the freedom of the air. They take their food and nourishment from the juice of wood and water during their mysterious and remarkable generation. I myself have seen many times and with my own eyes more than a thousand of these small bird-like creatures hanging from a single log upon the sea-shore. They were in their shells and already formed. No eggs are laid as is usual as a result of mating. No bird ever sits upon eggs to hatch them and in no corner of the land will you see them breeding or building nests. Accordingly in some parts of Ireland bishops and religious men eat them without sin during a fasting time, regarding them as not being flesh, since they were not born of flesh.

Pause, unhappy Jew! Pause – even if it be late. You hesitate to deny the first generation of man, from the slime of the earth without the co-operation of either man or woman; or the second, from man without the co-operation of woman – this, because of your veneration for the Law. The third only, that achieved by the co-operation of man and woman, because it is usual, you, with your hard neck, approve of and affirm. But the fourth generation, in which alone is salvation, that is from a woman without the co-operation of a man, you cannot, in your obstinate will, abide – and to your own destruction. Blush! wretch. Blush! At least consider the evidence of nature. She daily produces and brings forth new creatures without the co-operation of any male or female for our instruction and in confirmation of the Faith. The first generation was from slime; and this last from wood. The first indeed, because it happened only once through the operation of the Lord of nature, will ever seem worthy of all awe. But the last, though not less remarkable, provokes less wonder, because nature, who imitates, often produces it. For human nature is so made that only what is unusual and infrequent excites wonder or is regarded as of value. We make no wonder of the rising and the setting of the sun which we see every day; and yet there is nothing in the universe more beautiful or more worthy of wonder. When, however, an eclipse of the sun takes place, everyone is amazed – because it happens rarely.

12 ¶ *Birds that belong to two species and their natures*

THERE ARE MANY BIRDS HERE of a twofold nature. They are called ospreys. They are smaller than the eagle, but larger than the hawk. One of their feet is armed with talons, open and ready to snatch; but the other is closed and peaceful and suitable only for swimming. It is a wonderful instance of nature's pranks.

There is a remarkable thing about these birds, and I have often witnessed it for myself. They hover quietly on their wings high up in the air over the waves of the sea. In this way they can more easily see down into the depths below. Then, seeing with their sharp eyes through such a great distance of air and troubled water little fishes hiding below the waves, they dive down with amazing speed. While they enter and leave the water they control themselves by their swimming foot; but with their grasping foot they catch and carry off their prey.

So does our old enemy see with his sharp glance whatever secret thing we do in the troubled waves of this world. And while with peaceful claw he approaches us airily through success in temporal things, still with grasping claw, bloody with its booty, he seizes and destroys our unhappy souls.

13 ¶ *Birds that are so to speak mixed and not true to type*

ONE SHOULD REMARK that amongst both kinds of birds, ospreys and barnacles, some are found to be very like birds, but are, so to speak, deceptively and not truly so. They share the common nature of birds, and are far from having the characteristics of either type.

But the careful observing mind will clearly grasp the differences that exist between similar things and the similarities between different things.

14 ¶ *Kingfishers*[9] *and their natures*

HERE ARE FOUND ALSO those little birds which they call kingfishers. They are smaller than the blackbird, are rare, and are found on rivers. They are short like quails. They dive into the water in pursuit of very small fish on which they feed. While in all other respects they follow the nature of their type, here they are different in colour, but in that only. For they have a white belly and a black back. Elsewhere they are conspicuous in having a red belly, red beak and claws, and with bright shining green wings and back like the parrot or peacock.

There is a remarkable thing about these birds: if, when dead, they are kept in a dry place, they never putrefy; and if they are placed between clothes or anything else, they keep them free from the moth, and impart a pleasant perfume to them. And a thing even more amazing: if they are hung by the beak in a dry place, they change their coat of feathers each year, as if by virtue of a vital spirit that survives and continues to persist in some hidden part.

In the same way holy men, dead to the world, inflamed with love, and as it were left in a dry place, put on incorruption of life, and purify and perfect themselves and those in association with them from the moth of vice by their practice of good actions, and make them deserving of all praise by the good odour of their virtues. And while they hang from above, as it were, by their concentration of mind, they are daily renewed and changed for the better in the rejection of their old habits and the acquisition of virtues. They put off the old man entirely, and put on the new. For the advancement that is most deserving of praise is that in which what precedes is put into the second place by what follows.

15 ¶ *Swans and storks and their natures*

SWANS ARE VERY PLENTIFUL in the northern part of Ireland. Storks are very seldom seen anywhere in Ireland, and when they are, they are black.

There is a remarkable thing about swans. They teach us that the troubles of death should not grieve us; for in the very moment of dying they make a virtue of necessity and despise their sad fate in singing.[10]

So men, clothed in the white garments of virtue, depart joyfully from the hardships of this world. They thirst only for God, the living fountain, and desire to be freed and liberated from the body of this death and to be with Christ.

There is a remarkable thing about storks. They seek out waters that are ice-cold, and leave those that are warm. They spend the whole winter in the beds of rivers, but at the first sign of the mildness of spring, they emerge once again and take themselves off into the freedom of the air.

So holy men, during the winter of the world, sleep in the depths of the earth; but when the world is reformed to a better state and acquires an everlasting spring, they emerge at the first sound of the archangel, and being borne aloft in the air to meet Christ, and being called to his right hand, are transferred to the true liberty of the sons of God.

16 ¶ *Birds that do not appear in the winter-time*

IT IS REMARKABLE about these birds, and others of a like nature that are accustomed to disappear during the winter, that in the interval, neither dead nor alive, they seem to continue living in their vital spirit and at the same time to be seized up into a long ecstasy and some middle state between life and death. They receive no support from food by which the animal body is usually sustained, but, refreshed during such a long time by some hidden help of nature, suffer no bodily harm, and, wakened as it were from sleep, return with the 'zephyr' and 'the first swallow'.[11]

Like to theirs is the ecstasy of those whose souls are by divine permission rapt for some time to the highest dwelling places of the heavens or to the contemplation of the nether world. Eventually, having completed their mission, they return to their bodies, which in the meantime had been left on the earth, in a state worthy of all wonder, breathing without breath, living without life, and fully subject neither to life nor death.

17 ¶ *The many kinds of crows here and their natures*

THERE ARE NO BLACK CROWS in this country, or there are very few. What there are, are of different colours. These birds bring up small shell-fish into the air, and let them fall again so that they may be able to break by collision with a stone after a long fall the shell which they cannot break with their beak.

And so the old enemy with practised guile, as soon as he has advanced them to the highest peaks of honour, more boldly attacks those whom he could not pervert when they were in a lowly state. For driving them on by neglect of their duty, or, as they waver,

through the swelling of pride, he crushes and breaks them the more with a fall into the lowest depths of vice. The fall is greater in as much as they topple from a higher state.

There is a remarkable thing about these birds, for, although in other respects they are most cunning, nature has deprived them of sense only in the placing of their nests – a point in which even foolish birds are found to be clever. For building their nests in the public road, or in any gathering-place of men, or on a fallen tree, or on a rock, they make no provision against the winds, nor fear the approach of men or snakes.

And so every wise man in proportion to the greatness of his capacity or gift of wisdom with which he is adorned, if he be given to lust and be caught in the snares of passion, will in that proportion the less observe temperance and modesty. David and Solomon are good examples of this. A violent passion for silly women led one of them into the crime of homicide, and the other into the crime of apostasy.

Ireland has none but the best breed of falcons. Those inferior falcons, commonly called by the name *lanner*, are absent. Gerfalcons, which the northern and arctic regions beget and export, are absent. Partridges and pheasants are absent.[12] There are no magpies, and no nightingales.

18 ¶ *Wild animals and their natures and those that are missing*

IRELAND HAS almost all the kinds of wild animals that are found in the western regions. She has stags that are not able to escape because of their too great fatness. The smaller they are physically, the more nobly they carry themselves with the splendour of their heads and antlers. We have never seen anywhere such a supply of boars and wild pigs. But they are small, badly formed, and inclined to run away. They are equally inferior in their want of boldness and courage as in their physical make-up. There are many hares but rather small, and very like rabbits both in size and in the softness of their fur.

To put it briefly: you will find the bodies of all animals, wild-beasts, and birds smaller in their species than anywhere else. Only men retain their full size. There is a remarkable thing about those hares: if they are put up by dogs, they always try (unlike other hares) to make their escape in cover, as does the fox – in hidden country, and not in the open. They never make for the plains or rocky paths, unless they are driven and compelled to it. Martens are very common in the woods. They are hunted all day and all night, by means of fire.

Animals, except men, smaller than anywhere else.

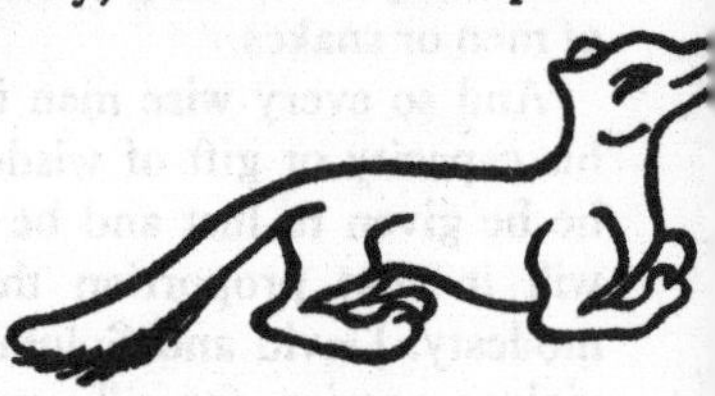

19 ¶ *The badger and its nature*

THE BADGER, or melot, is also found here. It is an unclean animal and tends to bite, frequenting rocky and mountainous places. Scraping and digging with its feet it makes for itself holes under the ground as places of refuge and defence. Some of them are born to serve by nature. Lying on their backs, they pile on their bellies soil that has been dug by others. Then clutching it with their four feet, and holding a piece of wood across their mouths, they are dragged out of the holes with their burdens by others who pull backwards while holding on here and there to the wood with their teeth. Anyone that sees them is astonished.

20 ¶ *The beaver and its nature*

BEAVERS USE a similar contrivance of nature. When they are building their homes in the rivers, they use slaves of their own kind as carts, and so by this wonderful means of transport pull and drag lengths of wood from the forests to the waters. In both kinds of animal (beaver and badger) the slaves are distinguished by a certain inferiority of shape and a worn bare patch upon their backs.

Ireland has badgers but not beavers. In Wales beavers are to be found only in the Teifi river near Cardigan. They are, in the same way, scarce in Scotland.

One should remark that beavers have wide tails, spread out like the palm of the human hand, and not long. They use them as oars in swimming. And while the whole of the rest of their body is very furry, they are entirely free from fur on this part, and are quite bare and slippery like a seal. Consequently in Germany and the northern regions, where beavers are plentiful, great and holy men eat the tails of beavers during fasting times – as being fish, since, as they say, they partake of the nature of fish both in taste and colour.

But about these and their nature, how and with what skill they build their settlements in the middle of rivers, and how, when pressed by an enemy, by the loss of a part they save the whole – a contrivance most commendable in an animal – will be more fully explained when we come to deal with the geography and description of Wales and Scotland, and the origin and nature of the people of each. We shall find another opportunity of doing this, and to another purpose, with God's help and if life be spared.

The island suffers the absence of certain wild beasts. Wild-goats, deer generally, hedgehogs and polecats are wanting. Moles are either not here at all, or at any rate, are very scarce on account of the great humidity of the country. Mice are infinite in number and consume much more grain than anywhere else, as well as eating garments even though they be locked up carefully. Bede says that there are

only two kinds of harmful beasts in Ireland, namely, wolves and foxes. I would add the mouse as a third, and say that it was very harmful indeed.

21 ¶ *Reptiles and those that are missing and how all poisonous reptiles are not found here*

OF ALL KINDS OF REPTILES only those that are not harmful are found in Ireland. It has no poisonous reptiles. It has no serpents or snakes, toads or frogs, tortoises or scorpions. It has no dragons. It has, however, spiders, leeches and lizards – but they are entirely harmless.

Some indulge in the pleasant conjecture that Saint Patrick and other saints of the land purged the island of all harmful animals. But it is more probable that from the earliest times, and long before the laying of the foundations of the Faith, the island was naturally without these as well as other things.[13]

22 ¶ *Poisonous reptiles when brought here immediately die and poison loses its force*

I DO NOT THINK it remarkable that the country should not have these reptiles, just as it has not got certain fish, birds, and wild beasts. It is a natural deficiency. But this fact is truly astonishing, namely, that if a poisonous thing is brought here from elsewhere, the island cannot, and never could, endure to keep it.

One reads in the ancient writings of the saints of that land that sometimes, by way of a test, snakes have been imported in bronze containers. But as soon as they reached the middle of the Irish Sea they were found to be lifeless and dead. And similarly if poison is brought in, it loses its natural force in the middle of its voyage through the operation of a kindly breeze. I have heard merchants that ply their trade on the seas say that sometimes, when they had unloaded their cargoes at an Irish port, they found toads brought in by chance in the bottom of the holds. They threw them out still living on to the land; but immediately they turned their bellies up, burst in the middle, and died, while everybody saw and wondered.

It is clear then, that, whether because of a clemency in the air that is, indeed, something new and never heard of before, but is nevertheless benign, or some hidden force of the land itself that is inimical to poisons, no poisonous animal can live here. And if poison be brought in, no matter what it be, from elsewhere, immediately it loses all the force of its evil.

23 ¶ *The dust of this land which kills poisonous reptiles*

INDEED the soil of this land is so inimical to poison that, if gardens or any other places of other countries are sprinkled with it, it drives all poisonous reptiles far away.

24 ¶ *The boot-thongs of this country used as antidotes against poisons*

THE THONGS also of that country, those really made from the hides of animals bred in the country, and not those merely said to be such, are wont to be an effective remedy, when cut up in little pieces and drunk with water, against the bites of serpents and toads.

I saw with my own eyes a thong of this kind placed in a complete circle around a toad to see what would happen. The toad came to it and wanted to pass, but, as if struck on the head, he fell back. He tried the opposite side, and then finding the thong everywhere, and fleeing it like a pest, in the end he suddenly dug a hole in the middle of the circle with his feet, and before the gaze of all went under the ground.

25 ¶ *A frog discovered in Ireland*

NEVERTHELESS in our days a frog was found near Waterford in some grassy land, and was brought to Robert Poer, who was in charge there at the time, and many others in assembly, both English and Irish, while it was still living. While the English, and more so the Irish, regarded it with great wonder, Duvenaldus, the king of Ossory, who happened to be there at the time, with a great shaking of his head and great sorrow in his heart at last said (and he was a man of great wisdom among his people and loyal to them): 'That reptile brings very bad news to Ireland.'

And regarding it as an indication of what was to be, he said that it was a sure sign of the coming of the English, and the imminent conquest and defeat of his people.

No one will suppose that the frog had ever been born in Ireland; because here the mud does not, as elsewhere, 'contain the seeds from which green frogs are born'.[14] For if it had been otherwise, frogs would frequently and in numbers have been found either before or after this. But perhaps it came from some neighbouring port in a ship by chance, was thrown out on the land, and, since it was not a poisonous reptile, was able to live for some time.

26 ¶ *The many good points of the island and the natural qualities of the country*

THIS IS the most temperate of all countries. Cancer does not here drive you to take shade from its burning heat; nor does the cold of Capricorn send you rushing to the fire. You will seldom see snow here, and then it lasts only for a short time. But cold weather does come with all the winds here, not only from the west-north-west and north but also equally from the east, the Favonius and the Zephyr. Nevertheless, they are all moderate winds and none of them is too strong. The grass is green in the fields in winter, just the same as in summer. Consequently the meadows are not cut for fodder, nor do they ever build stalls for their beasts. The country enjoys the freshness and mildness of spring almost all the year round.

Cold weather comes with all the winds here.

The air is so healthy that there is no disease-bearing cloud, or pestilential vapour, or corrupting breeze. The island has little use for doctors. You will not find many sick men, except those that are actually at the point of death. There is here scarcely any mean between constant health and final death. Anyone born here, who has never left its healthy soil and air, if he be of the native people, never suffers from any of the three kinds of fevers. They suffer only from the ague and even that only very seldom.

The island has little use for doctors.

This indeed was the true course of nature; but as the world began to grow old, and, as it were, began to slip into the decrepitude of old age, and to come to the end, the nature of almost all things became corrupted and changed for the worse.

There is, however, such a plentiful supply of rain, such an ever-present overhanging of clouds and fog, that you will scarcely see even in the summer three consecutive days of really fine weather. Nevertheless, there is no disturbance of the air or inclemency of the weather such as inconveniences those that are in health and spirits, or distresses those that suffer from nervous disorders.

27 ¶ *The advantages of the West are to be preferred to those of the East*

WHAT RICHES has the East then to offer in comparison with these? It has, of course, many-coloured silken cloth produced by the silk-worm; it has precious metals of certain types, sparkling gems and aromatic spices. But what are these in comparison with the loss of life and health? They are obtained only by enduring constantly the enmity of an enemy that one cannot get away from – the air that is within, and that surrounds one.

28 ¶ *All the elements in the East are pestiferous*

ALL THE ELEMENTS in the East, even though they were created for the help of man, threaten his wretched life, deprive him of health, and finally kill him. If you put your naked foot upon the ground, death is upon you; if you sit upon marble without taking care, death is upon you; if you drink unmixed water, or merely smell dirty water with your nostrils, death is upon you; if you uncover your head to feel the breeze the better, it may affect you by either its heat or coldness – but in any case, death is upon you. The heavens terrify you with their thunder and threaten you with their lightnings. The sun with its burning rays makes you uncomfortable. And if you take more food than is right, death is at the gate; if you take your wine unmixed with water, death is at the gate; if you do not hold back your hand from food long before you are satisfied, death is at the gate.

29 ¶ *The venom of poisons there*

THE POISONED HAND is to be feared there too: that of his step-mother by the step-son, of the enraged wife by her husband, and that of his wicked cook by the master. And not only food and drink, but also clothes, chairs and seats of all kinds. Poisons attack you from all sides, as also do poisonous animals. But the most harmful of all harmful things attacks you – man himself. Among so many dangers of death, what feeling can there be of security of life? Or rather, among so many deaths, what life can there be?

30 ¶ *The incomparable mildness of our climate*

LET THE EAST, then, have its riches – tainted and poisoned as they are. The mildness of our climate alone makes up to us for all the wealth of the East, in as much as we possess the golden mean in all things, giving us enough for our uses and what is demanded by nature.

O gift from God, on this earth incomparable! O grace, divinely bestowed on mortals, inestimable, and not yet appreciated! We can safely take our rest in the open air, or upon bare marble. We have no fear of any breeze, piercing in its coldness, fever-laden with its heat, or pestilential in what it brings. The air, that by breathing in we encompass and which continually encompasses us, is guaranteed to us to be kindly and health-giving.

31 ¶ *Certain deficiencies here that are in fact praiseworthy*

THERE ARE certain other things also which, as in the case of reptiles, are wanting here, but whose absence is a good thing. There never are earthquakes here. You will hear thunder here scarcely once in the year. Accordingly thunder does not frighten, nor lightning terrify one. No cataracts rush down upon one. No earthquakes swallow one. The lion does not prowl, nor the panther tear to pieces, nor the bear devour, nor the tiger eat one up. No hospitality is dangerous because of the suspicion of poison – even in the case of an enemy. The step-son fears no poisoned cup from his step-mother, nor the husband from his enraged wife.

32 ¶ *The well of poisons is in the East*

THE WELL OF POISONS brims over in the East. The farther therefore from the East it operates, the less does it exercise the force of its natural efficacy. And by the time it reaches these farthest parts, after having traversed such long distances, losing its force gradually, it is entirely exhausted – just as the sun, the farther from the zodiac it sends its rays, the less does it exercise the force of its heat; and so some of the farthest parts of the arctic regions are completely without the benefit of its warmth.

Objection.

But you may object that the East excels in precious stones and roots with certain virtues.

Reply.

Nature has indeed provided that where there were many evils, there should be many remedies against those evils. For the experience of many diseases compels one to seek for many medicines. But here, where dangers to health are so very few, so too are the remedies. By so much, then, as peace is more desirable than anxiety, preservation than cure, the satisfaction of health constantly enjoyed to the searching after remedies for serious illnesses, so the advantages of the West outshine and outstrip those of the East, and nature has given a more indulgent eye to the regions traversed by the west wind than those traversed by the east.

¶ Here begins the second part of the History of Ireland

33 ¶ *The Wonders*[15] *and Miracles of Ireland*

NOW WE TURN OUR PEN TO THOSE things which, appearing to be contrary to nature's course, are worthy of wonder.

I have thought it worth while to give some account of such things as are marvellous in themselves and, because of their recent origin, are easily seen and have been placed in these parts by nature herself. I have also treated of those exceedingly wonderful and miraculous deeds done through the merits of the saints. They are worthy of being set on record for ever in view of the testimony on which they are received. For just as the marvels of the East have through the work of certain authors come to the light of public notice, so the marvels of the West which, so far, have remained hidden away and almost unknown, may eventually find in me one to make them known even in these later days.

I am aware that I shall describe some things that will seem to the reader to be either impossible or ridiculous. But I protest solemnly that I have put down nothing in this book the truth of which I have not found out either by the testimony of my own eyes, or that of reliable men found worthy of credence and coming from the districts in which the events took place.

And it should not seem surprising if wonderful things are written about the works of him who made whatever he wished. For 'God is wonderful in his saints', and 'great in all his works'. And the psalmist is elsewhere made to say: 'Come and see the works of the Lord, the wonders that he has worked on the earth.' The careful reader will know that history does not spare the truth, and recounts rather what is in fact true than that which seems like the truth.

34 ¶ *The surging tides of the Irish Sea and its varying ebb and flow*

THE IRISH SEA, surging with currents that rush together, is nearly always tempestuous, so that even in the summer it scarcely shows itself calm for a few days to them that sail.[16]

35 ¶ *The contrary motions of the tides in Ireland and Britain*

AS OFTEN AS THE WATER EBBS to half depth from the port of Dublin, it flows to a half depth into the port of Milford, in Britain – a very good harbour for ships. Gradually the water flows back into the farther shores of Bristol that had been laid bare by the retreating tide. The same happens the other way round when the tide is flowing.

There is a harbour at Wicklow, on that side of Ireland that, being nearest, looks towards Wales, which fills up when the tide is ebbing from all other places. But, when the tide is flowing elsewhere, it is left dry. There is another notable thing about this place: when the sea has ebbed and the whole shore is dry, a river flowing in is made bitter throughout its whole channel by a continual saltiness. But the opposite happens in the nearby harbour at Arklow. There the river that flows into the sea at that point retains its natural sweetness without any diminishing, not only when the tide has ebbed completely, but also when the tide is flowing in and filling the whole bay. In this way it keeps its waters free from the salt until they reach the sea itself.

36 ¶ *The moon affects the waters as well as natural humours*

WHEN THE MOON is at her meridian, the ocean recalling its attendant waters to its hidden lairs always leaves the coasts of Britain entirely dry. In the meantime it fills up the Irish coasts near Dublin with the ebbing tide. The coast near Wexford follows the flow, not of the Irish tides at Dublin, but of the British tides at Milford. And a thing which is even more to be wondered at: there is a certain rock out in the sea not far from Arklow from which the tide ebbs on one side, while it flows in on the other.

When the moon is recovering her light, and is already beginning to grow beyond her half size and becoming big, the western seas through some secret natural causes begin to be rough and tempestuous; they swell more and more from day to day with surging waves running up on the shore in full spate far beyond their usual limits, until the moon has attained the full perfection of her roundness. But as her fires decrease and she turns her face away, the swelling of the waves begins to abate until, when she fails altogether, the tide loses its flood and returns once again to its usual channels.

Indeed Phoebe is to such an extent a source and influence on all liquids, that according to her waxing and waning she directs and controls not only the waves of the sea, but also the bone-marrow and brains in all living things as well as the sap of trees and plants. When she is deprived of her full light you will notice that all things lose their fullness. But when she has attained her complete roundness, you will find that bones are full of marrow, heads of brains, and other things of sap.

But it would take a greater work than this to explain the reasons behind such things, and why the western seas have taken for themselves in some well-ordered and unfailing liveliness this surging and falling, rather than or more clearly than the eastern seas; and why all these things happen at the direction of the moon, the controller of things liquid. I have, in fact, explained these points clearly and briefly in a book written in verse and called *The Flowers of Philosophy*.[17]

37 ¶ *Two islands in one of which no one dies; into the other no animal of the female sex can go*[18]

THERE IS A LAKE in the north of Munster which contains two islands, one rather large and the other rather small. The larger has a church venerated from the earliest times. The smaller has a chapel cared for most devotedly by a few celibates called 'heaven-worshippers' or 'god-worshippers'.[19]

No woman or animal of the female sex could ever enter the larger island without dying immediately. This has been proved many times by instances of dogs and cats and other animals of the female sex. When brought there often to make a trial, they immediately died.

A remarkable thing about the birds there is that, while the males settle on the bushes everywhere throughout the island, the females fly over and leave their mates there and, as if they were fully conscious of its peculiar power, avoid the island like a plague.

In the smaller island no one ever died or could die a natural death. Accordingly it is called the island of the living. Nevertheless the inhabitants sometimes suffer mortal sicknesses and endure the agony almost to their last gasp. When there is no hope left; when they feel that they have not a spark of life left; when as the strength decreases they are eventually so distressed that they prefer to die in death than drag out a life of death, they get themselves finally transported in a boat to the larger island, and, as soon as they touch ground there, they give up the ghost.

38 ¶ *An island one part of which is frequented by good and the other by evil spirits*[20]

THERE IS A LAKE in Ulster which contains an island divided into two parts.

One part contains a very beautiful church with a great reputation for holiness, and is well worth seeing. It is distinguished above all other churches by the visitation of angels and the visible and frequent presence of local saints.

But the other part of the island is stony and ugly and is abandoned to the use of evil spirits only. It is nearly always the scene of gatherings and processions of evil spirits, plain to be seen by all. There are nine pits in that part, and if anyone by any chance should venture to spend the night in any one of them – and there is evidence that some rash persons have at times attempted to do so – he is seized immediately by malignant spirits, and is crucified all night with such severe torments, and so continuously afflicted with many unspeakable punishments of fire and water and other things, that, when morning comes, there is found in his poor body scarcely even the smallest trace of life surviving. They say that if a person once undergoes these torments because of a penance imposed on him, he will not have to endure the pains of hell – unless he commit some very serious sin.

39 ¶ *An island*[21] *where human corpses exposed in the open do not putrefy*

THERE IS AN ISLAND in the sea west of Connacht which is said to have been consecrated by Saint Brendan. In this island human corpses are not buried and do not putrefy, but are placed in the open and remain without corruption. Here men see with some wonder and recognize their grandfathers, great-grandfathers, and great-great-grandfathers and a long line of ancestors.

There is another remarkable thing about this island: while the whole of Ireland is infested with mice, there is not a single mouse here. For no mouse is bred here nor does one live if it be brought in. If by chance it is brought in, it makes straight for the nearest point of the sea and throws itself in; and if it be prevented, it dies on the spot.

40 ¶ *The wonderful natures of wells*

THERE IS A WELL in Munster[22] and if anyone washes in its waters, he immediately turns grey. I saw a man who had washed there one part of his beard. It had turned grey, while the other part retained its natural dark colour. On the other hand there is a well in Leinster and if a man washes in it, he will not get greyer. There is a well of sweet water in Connacht on the top of a high mountain and some distance from the sea, which in any one day ebbs and overflows three times, imitating the ebbing and flowing of the sea. In Wales beside the castle of Dinevor there is a well whose waters are similarly inconstant. There is a well in the far north of Ulster which is so cold that if logs of wood are left in it for seven years, they harden so as to become stones.

There is in Norway also a well of the same kind, but its efficacy is the greater in as much as it is nearer to the frigid zone. Not only if wood, but if flax or a linen web be placed in it for a year, it becomes very hard stone. Because of this a certain bishop of Norway was able to bring back to Waldemar, king of Denmark in our time, a napkin which he had taken from him the year before for the purpose of proving the fact referred to. The napkin had now a double nature. The middle part, having been placed in the well, was a stone, but the rest of it remained as before, having been outside of the well.

While it is generally accepted that the use of the elements is naturally common to all living things, there is in Connacht a well which has water that can be drunk only by men, but is pestilential to horses or cattle, or any other animals whatever that should taste it. Little pebbles taken from this well and placed in the mouth quench the thirst when one is dry and parched.

There is a well in Munster,[23] and if one touches or even looks at it, the whole province is deluged with rain. The rain will not cease until a priest, who is a virgin both in mind and body and specially chosen for the purpose, celebrates Mass in a chapel not far from the well and known to have been erected with this end in view, and appeases the well with a sprinkling of holy water and the milk of a cow of one colour. This is certainly a barbarous rite, without rime or reason.

41 ¶ *Two wonderful wells, one in Brittany and the other in Sicily*

THERE IS A WELL in Brittany of a similar nature to the one preceding. If you drink its water out of the horn of an ox and happen to spill any of it on a nearby rock, a shower of rain will fall on you immediately, even though the sky be ever so clear from rain.

There is in Sicily a well of a wonderful nature. If anyone approaches it clad in a red garment, a column of water immediately shoots up from the well to the height of the person. This well is entirely unmoved by any other colour, and if the person in red departs, it resumes its ordinary dimensions, and returns to its former channels.

'Happy indeed is he who is able to know the causes of things.'[24] But since there is a limit to human capacity, and whatever is human is far from being perfect, 'do ye, O Muses, tell' the causes of such things as these. 'We cannot all do all.'[25] Envious nature has locked and hidden away for herself among her secret wonders the causes of these and other such things.

There is on the sea-shore in Connacht a rocky mound, fairly big, which seems to tower as much above the waves when the tide returns and covers even high objects as it does over the shore when the tide has ebbed. There is also in Connacht a certain walled place that looks like a large castle, and is said to have been consecrated by Saint Patrick. Although the herds of the whole province are often placed there for refuge, there is never such a number there that many more cannot be put in, until somebody happens to say that there is enough in, or expresses some doubt as to its being able to hold more.

42 ¶ *A big lake[26] that had a marvellous origin*

THERE IS A LAKE in Ulster of a remarkable size. It is thirty miles long and fifteen miles wide. From it a very beautiful river called the Bann flows into the northern ocean. Here the fishermen complain not of a scarcity of fish, but of too great catches and the breaking of their nets. In our time a fish was caught here – one that had come down from the lake, and not from the sea – which had the shape more or less of a salmon, and was of such size that it could not be dragged or carried as a whole. Accordingly it was cut up and carried about through the province.

They say that an accident was responsible for the rise of this remarkable lake. There was from ancient times in the region now covered by the lake a people very much given to vice, and particularly addicted, above any other people in Ireland, to bestiality. There was a saying well known to that people that if a certain well of the district which, because of a great fear of it that had been inherited from a barbarous superstition, was always covered and sealed, should be left uncovered, it would immediately overflow to such an extent that it would wipe out and destroy the whole district and people. It happened, however, that a young woman came to the well to draw water. She filled her vessel, and, before covering the well, ran quickly to her little child, because she had heard him crying where she had placed him a little way off. But 'the voice of the people is the voice of God', and when she hurried back she met such an overflow from the well that both herself and her boy were swept off immediately. Within an hour the whole people and their flocks were overwhelmed in this local and provincial flood. The whole area was covered with a sea of water which remained there and made a permanent lake.

It looked as if the author of nature had judged that a land which had known such filthy crimes against nature was unworthy not only of its first inhabitants but of any others in the future.

There is some confirmation of this story in as much as fishermen of the lake clearly see under the waves in calm weather towers of churches, which, as is usual in that country, are tall, slender, and rounded. They frequently point them out to visitors who are amazed at the occurrence. One should note, however, that the river which now flows from the lake and is of such great size, always, from the time of Bartholanus after the Flood, traced its course back to the aforementioned well, and, joined by other rivers, flowed through the same country, but was of a much smaller size. It was, in fact, one of the nine principal rivers.

43 ¶ *A fish with three gold teeth*

TWO YEARS before the coming of the English to the island,[27] there was found at Carlingford in Ulster a fish of unusual size and quality. Among other wonderful things about it was that it had three teeth of gold of about fifty ounces' weight in all. It seemed to prefigure the imminent conquest of the country.

44 ¶ *The northern islands, almost all of which are held by the Norwegians*

IN THE NORTHERN OCEAN beyond Ulster and Galloway are a number of islands, namely the Orcades and the Incades, and many others. Almost all of them are held by, and are subject to, the Norwegians. For even though they lie much nearer to other regions, nevertheless the Norwegians, who keep their eyes ever on the ocean, lead, above any other people, a piratical life. Consequently all their expeditions and wars are decided by naval engagements.

45 ¶ *The island that was at first unstable and then was made stable by means of fire*

AMONG THE OTHER ISLANDS is one that arose recently, and which they call the 'phantom' island. Its origin came about in this way.

One fine day the inhabitants of the islands noticed that a large mound of earth arose in the sea where land had never been seen before. They all wondered. Some said that it was a whale or some other monstrous sea animal. Others, however, reflecting that it remained without any movement, said 'No, not at all. It is land.'

They wanted, nevertheless, to remove the uncertainty, and so they selected young men from the island nearest to it, and sent them over to it in a rowing boat. When they had approached so near that they were on the point of disembarking, the island disappeared entirely from before their eyes as if descending into the sea. It appeared again in the same way on the following day, only, however, to play a similar trick on the same young men. On the third day they followed the advice of a certain old man in first throwing at the island as they approached it an arrow of red-hot iron, and then disembarking, they found the land stable and habitable.

And so there are many proofs that fire is always most hostile to phantoms. Hence those who have seen a phantom cannot look upon the splendour of fire without falling into a swoon immediately. For fire by its position and nature is the most noble of the elements. It is, as it were, aware of the secrets of the heavens. The sky is of fire; the planets are of fire; the bush was on fire but was not burned. The Holy Spirit came upon the apostles with tongues of fire.

46 ¶ *Iceland whose people speak few words but true and never swear an oath*

ICELAND, the largest of the islands of the north, lies at a distance of three days' sailing to the north of Ireland. Its people say little but they always tell the truth. They speak but seldom and briefly and never use an oath. They do not know how to lie. They detest nothing more than a lie. Their priest is their king, and their king is their priest. The bishop has the powers of both kingship and priesthood. This land produces and rears gerfalcons and big and noble hawks.

47 ¶ *A whirlpool of the sea that swallows ships*[28]

NOT FAR TO THE NORTH from the islands of which we have been speaking is a certain wonderful whirlpool in the sea. All the waves of the sea from all parts, even those remote, flow and strike together here as if by agreement. They seem to pour themselves into the secret recesses of nature here, and are, so to speak, devoured in the abyss. If a ship happens to touch it, it is caught and pulled with such violence of the waves that the force of the pull downwards immediately swallows it up for ever.

48 ¶ *The Isle of Man which on account of its admission of poisonous reptiles is judged to belong to Britain*

AMONG THE SMALLER ISLANDS there is one of fair size that is now called the Isle of Man, but in antiquity was called Ewania. They say that it is equidistant from the north of Ireland and Britain. There was a great controversy in antiquity concerning the question: to which of the two countries should the island properly belong? Eventually, however, the matter was settled. All agreed that since it allowed poisonous reptiles to live in it, it should belong to Britain.

49 ¶ *That a long time after the Flood and then not suddenly but gradually and as it were through inundation islands came to be*

WHETHER ISLANDS CAME TO BE before the Flood or because of the Flood, when all the parents of animals were enclosed in the ark, it is nevertheless very difficult to answer why harmful animals, and among them poisonous reptiles, came to fill the remoter islands. It is clear that no one in his senses would have brought them there. On this point we can say with some probability that it was long after the Flood, when the earth was already full everywhere of the animals that had multiplied, that the islands were formed, and then not suddenly, or violently, but gradually and as it were through inundation.

50 ¶ *Thule an island of the West very well known in the East but entirely unknown in the West*

A REMARKABLE THING about Thule, which is said to be the farthest of the western islands, is that it is very well known among the eastern people both in name and for its nature, although it is entirely unknown to the people of the West. Solinus says[29] that it is the farthest island among the many around Britain. He says that there, during the summer solstice, there is no night, and, during the winter solstice, there is no day.

But whatever may be said of its name, it is quite clear that none of the western islands has such a nature. In the farthest parts, however, of the northern region, the sun turning round from Cancer seems to revolve around the edge of the earth, but over the horizon, for the space of a few nights. And when it is returning from the constellation of Capricorn, the brightness of its longed-for light disappears, as it were, beneath the dark limits of the Antarctic pole for the space of the same number of days.

51 ¶ *The Giants' Dance which was transferred from Ireland to Britain*

THERE WAS in ancient times in Ireland a remarkable pile of stones which was called the Giants' Dance, because giants brought it from the farthest limits of Africa to Ireland, and erected it, according to some on a mountain of Kildare,[30] according to others at Ophela near the castle of Nas, employing truly remarkable skill and ability. It is amazing how so many great stones were ever brought together or erected in one place, and with what skill upon such great and high stones others no less great were placed. These latter seem to be hanging, as it were, and suspended in space, so as to rest rather on the skill of the craftsmen than on the base of stones beneath.

According to British history,[31] the king of the Britons, Aurelius Ambrosius, arranged through the divine help of Merlin that these stones be brought over from Ireland to Britain. He got them put up in exactly the same order and with the same skill as before – so as to leave behind some memorial of a great crime committed when the flower of Britain's manhood was cut to pieces by the concealed daggers of the Saxons, who, coming in the guise of peace with the weapons of treachery, killed the youth of the kingdom that had been so carelessly guarded.

52 ¶ *The wonderful happenings of our own time; and first about a wolf that talked with a priest*

AND NOW WE SHALL GIVE some account of the things that occurred in our own time and seem worthy of wonder.

About three years before the coming of Lord John into Ireland,[32] it happened that a priest, journeying from Ulster towards Meath, spent the night in a wood on the borders of Meath. He was staying up beside a fire which he had prepared for himself under the leafy branches of a tree, and had for company only a little boy, when a wolf came up to them and immediately broke into these words: 'Do not be afraid! Do not fear! Do not worry! There is nothing to fear!'

They were completely astounded and in great consternation. The wolf then said some things about God that seemed reasonable. The priest called on him and adjured him by the omnipotent God and faith in the Trinity not to harm them and to tell them what kind of creature he was, who, although in the form of a beast, could speak human words. The wolf gave a Catholic answer in all things and at length added:

'We are natives of Ossory. From there every seven years, because of the imprecation of a certain saint, namely the abbot Natalis, two persons, a man and a woman, are compelled to go into exile not only from their territory but also from their bodily shape. They put off the form of man completely and put on the form of wolf. When the seven years are up, and if they have survived, two others take their place in the same way, and the first pair return to their former country and nature.

'My companion in this pilgrimage is not far from here and is seriously ill. Please give her in her last hour the solace of the priesthood in bringing to her the revelation of the divine mercy.'

This is what he said, and the priest, full of fear, followed him as he went before him to a certain tree not far away. In the hollow of the tree the priest saw a she-wolf groaning and grieving like a human being, even though her appearance was that of a beast. As soon as she saw him she welcomed him in a human way, and then gave thanks also

to God that in her last hour he had granted her such consolation. She then received from the hands of the priest all the last rites duly performed up to the last communion. This too she eagerly requested, and implored him to complete his good act by giving her the viaticum. The priest insisted that he did not have it with him, but the wolf, who in the meantime had gone a little distance away, came back again and pointed out to him a little wallet, containing a manual and some consecrated hosts, which the priest according to the custom of his country carried about with him, hanging from his neck, on his travels. He begged him not to deny to them in any way the gift and help of God, destined for their aid by divine providence. To remove all doubt he pulled all the skin off the she-wolf from the head down to the navel, folding it back with his paw as if it were a hand. And immediately the shape of an old woman, clear to be seen, appeared. At that, the priest, more through terror than reason, communicated her as she had earnestly demanded, and she then devoutly received the sacrament. Afterwards the skin which had been removed by the he-wolf resumed its former position.

When all this had taken place – more in equity than with proper procedure – the wolf showed himself to them to be a man rather than a beast. He shared the fire with them during the whole of the night, and when morning came he led them over a great distance of the wood, and showed them the surest way on their journey. When they parted he gave many thanks to the priest for the benefit he had conferred upon him, and promised to give him much more tangible evidence of his gratitude, if the Lord should call him back from the exile in which he was, and of which he had now completed two thirds.

Almost two years later I happened to be passing through Meath where the bishop of that region had called a synod. He had also summoned the neighbouring bishops and abbots so that, advised by their counsel, he might more clearly see what he should do in the matter recounted and which he had learned on the confession of the priest. When he heard that I was going through those parts, he sent two of his clerics to me, asking me to come in person, if I could, to discuss so serious a matter. If, however, I could not come, I was at least to indicate my view in writing. When I had heard the whole account (which I had, as a matter of fact, heard already from others) in due order from them, and since I could not be present because of urgent business, I gave them the benefit at least of my advice in writing. The bishop and synod agreed with it, and sent the priest to the Pope with his documents, in which were given an account of the affair and the priest's confession, and which were sealed with the seals of the bishops and abbots that were present.

53 ¶ *A woman with a beard and a mane on her back*

DUVENALDUS, the king of Limerick, had a woman that had a beard down to her waist. She had also a crest from her neck down along her spine, like a one-year-old foal. It was covered with hair. This woman in spite of these two enormities was, nevertheless, not hermaphrodite, and was in other respects sufficiently feminine. She followed the court wherever it went, provoking laughs as well as wonder. She followed neither fatherland nor nature in having a hairy spine; but in wearing her beard long, she was following the custom of her fatherland, not of her nature.

54 ¶ *A man that was half an ox and an ox that was half a man*

IN THE NEIGHBOURHOOD of Wicklow at the time when Maurice fitzGerald got possession of that country and the castle,[33] an extraordinary man was seen – if indeed it be right to call him a man. He had all the parts of the human body except the extremities which were those of an ox. From the joinings of the hands with the arms and the feet with the legs, he had hooves the same as an ox. He had no hair on his head, but was disfigured with baldness both in front and behind. Here and there he had a little down instead of hair. His eyes were huge and were like those of an ox both in colour, and in being round. His face was flat as far as his mouth. Instead of a nose he had two holes to act as nostrils, but no protuberance. He could not speak at all; he could only low. He attended the court of Maurice for a long time. He came to dinner every day and, using his cleft hooves as hands, placed in his mouth whatever was given to him to eat. The Irish natives of the place, because the youths of the castle often taunted them with begetting such beings on cows, secretly killed him in the end in envy and malice – a fate which he did not deserve.

Shortly before the coming of the English into the island a cow from a man's intercourse with her – a particular vice of that people – gave birth to a man-calf in the mountains around Glendalough. From this you may believe that once again a man that was half an ox, and an ox that was half a man[34] was produced. It spent nearly a year with the other calves following its mother and feeding on her milk, and then, because it had more of the man than the beast, was transferred to the society of men.

55 ¶ *A cow that was partly a stag*

NEAR CHESTER in Britain a cow that was partly a stag was born in our time from the intercourse of a stag with a cow. All the fore parts as far as the groin were bovine, but the thighs and the tail, the hind legs and the feet, were clearly those of a stag, especially in quality and colour of hair. But since it was more of a cow than a wild animal it stayed with the herd.

56 ¶ *A goat that had intercourse with a woman*

ROTHERICUS, king of Connacht, had a tame white goat that was remarkable of its kind for the length of its coat and height of its horns. This goat had bestial intercourse with a certain woman to whom he was entrusted. The wretched woman, proving herself more a beast in accepting him than he did in acting, even submitted herself to his abuse.

How unworthy and unspeakable! How reason succumbs so outrageously to sensuality! That the lord of the brutes, losing the privileges of his high estate, should descend to the level of the brutes, when the rational submits itself to such shameful commerce with a brute animal! Although the matter was detestable on both sides and abominable, yet was it less so by far on the side of the brute who is subject to rational beings in all things, and because he was a brute and prepared to obey by very nature. He was, nevertheless, created not for abuse but for proper use. Perhaps we might say that nature makes known her indignation and repudiation of the act in verse:

> Only novelty pleases now: new pleasure is welcome;
> Natural love is outworn
> Nature pleases less than art; reason, no longer reasoning,
> Sinks in shame.

57 ¶ *A lion that loved a woman*

I SAW IN PARIS a lion which a cardinal had given when it was a whelp to Philip the son of Louis, then a boy. This lion used to make beastly love to a foolish woman called

Johanna. Sometimes when he escaped from his cage and was in such fierce anger that no one would dare to go near him, they would send for Johanna who would calm his anger and great rage immediately. Soothing him with a woman's tricks, she led him wherever she wanted and changed all his fury immediately into love.

O Beasts! Both! Worthy of a shameful death! But such crimes have been attempted not only in modern times but also in antiquity, which is praised for its greater innocence and simplicity. The ancients also were stained with such unspeakable deeds. And so it is written in Leviticus: 'If a woman approaches any beast to have intercourse with him, ye shall kill the woman, and let the beast die the death.'[35] The beast is ordered to be killed, not for the guilt, from which he is excused as being a beast, but to make the remembrance of the act a deterrent, calling to mind the terrible deed.

58 ¶ *How the cocks in Ireland crow differently from those in other countries*

DOMESTIC COCKS do not here, as elsewhere, distinguish the third and last part of the night from the two preceding by crowing three times at intervals. They are heard here rarely before daylight; and as elsewhere the day is known to be at hand from the third, so here it is known from the first crowing of the cock. Nevertheless it is not to be thought that the cocks have here a different nature from those of other places. For if they are brought here from outside they crow the same as the others here. But as Britain is content with a short night, so too is Ireland, and in fact it is shorter here, since Ireland is nearer the setting of the sun. And therefore, by how much shorter the night is here, so is the day nearer at hand after the crowing of the cock. Consequently during the summer-time the heavens above the horizon are always bright all night long from the brightness of the sun which scarcely dips under the land. One would think it was daylight with the dawn coming up.

59 ¶ *Wolves that whelp in December*

WOLVES IN IRELAND generally have their young in December, either because of the extreme mildness of the climate, or rather as a symbol of the evils of treachery and plunder which here blossom before their season.

60 ¶ *Ravens and owls that have their young about Christmas-time*

ABOUT CHRISTMAS-TIME of the year when Lord John first left the island,[36] ravens and owls in many parts of the country had their young. Perhaps they foretold the occurrence of some new and premature evil.

61 ¶ *About miracles; and first about the fruit[37] and ravens and blackbird of Saint Kevin*

NOW LET US COME to miracles, and let us begin with Saint Kevin, a great confessor of the Faith, and abbot.

At the time when Saint Kevin was distinguished for his life and sanctity at Glendalough, a noble boy, who was a student of his, happened to be sick and to ask for fruit. The saint had pity on him and prayed for him to the Lord. Whereupon a willow-tree not far from the church brought forth fruit that was health-giving to the boy and to others that were sick. And to this day both the willow-tree and others planted from it around the cemetery like a wall of willows bring forth fruit each year, although in all other respects, in their leaves and branches, they retain their own natural qualities. This fruit is white and oblong in shape, health-giving rather than pleasant to the taste. The local people have a great regard for it. Many bring it to the farthest parts of Ireland to cure various diseases, and it is called the fruit of Saint Kevin.

On the feast-day of the saint, the ravens at Glendalough, perhaps because they spilled the milk of the same student, are prevented by a curse of Saint Kevin from alighting on the earth or taking food. They fly around the village and the church, making a great noise, and on that day enjoy neither rest nor food.

Once upon a time the same Saint Kevin fleeing during Lent, as was his wont, the society of men, was by himself in a small cabin which warded off from him only the sun and the rain. He was giving his attention to contemplation and was reading and praying. According to his custom he put his hand, in raising it to heaven, out through the window, when, behold, a blackbird happened to settle on it, and using it as a nest, laid its eggs there. The saint was moved with such pity and was so patient with it that he neither closed nor withdrew his hand, but held it out in a suitable position without tiring until the young were completely hatched out. In perpetual remembrance of this wonderful happening, all the representations of Saint Kevin throughout Ireland have a blackbird in the outstretched hand.

62 ¶ *The teal of Saint Colman which are almost tame but fly away from hurt*

THERE IS A LAKE in Leinster that is not very big and is the haunt of the birds of Saint Colman. They are small ducks which are commonly called teal. Since the time of the saint these have become so tame that they take food from one's hand, and do not fear even to this day the approach of men. Whenever any injury or molestation happens to the church, the clergy, or themselves, they go off to a lake at some distance, and do not return to their former abode until due satisfaction has been made. In the meantime during their absence the waters of the lake, which before were limpid and clear, become brackish and dirty, and cannot be used either for man or beast.

It happened once that a person drawing water there by night drew out of the water (by chance, and not design) one of the birds. He was cooking meat for a long time in the water which he had drawn, and still it was not cooked nor could it be cooked. At length the bird was found swimming around in the vessel entirely unhurt. As soon as it was brought back to the lake, the meat was immediately cooked.

It happened in our own times that Robert fitzStephen was travelling through that place with Dermot the king of Leinster, when an archer struck one of those birds with an arrow. He took it off with him to his quarters and placed it in a pot with some other meat to be cooked. He burned out three fires and spent until midnight trying to cook it, and still he did not succeed. When he took out the meat for the third time, it was still found to be as raw as when he put it in. At length when his host saw the body of the little bird among the pieces of meat, and heard that it had been taken from that lake, he immediately exclaimed with tears:

'Alas for me! That this misfortune should ever have happened in my house! This was one of the birds of Saint Colman.'

And then the meat which could not be cooked, when put in the pot by itself, was immediately done. Not long afterwards the archer perished miserably.

63 ¶ *The stone which every day miraculously contains wine*

IN THE SOUTH of Munster near Cork there is a certain island which has within it a church of Saint Michael, revered for its true holiness from ancient times. There is a certain stone there outside of, but almost touching, the door of the church on the right-hand side. In a hollow of the upper part of this stone there is found every morning through the merits of the saints of the place as much wine as is necessary for the celebration of as many Masses as there are priests to say Mass on that day there.

64 ¶ *The fleas that were banished by Saint Nannan*

THERE IS IN CONNACHT a village celebrated for a church of Saint Nannan. In olden times there was such a multitude of fleas there that the place was almost abandoned because of the pestilence, and was left without inhabitants, until, through the intercession of Saint Nannan, the fleas were brought to a certain neighbouring meadow. The divine intervention because of the merits of the saint so cleansed the place that not a single flea could ever afterwards be found here. But the number of them in the meadow is so great that it ever remains inaccessible not only to men but also to beasts.

65 ¶ *The rats that were expelled from Ferneginan[38] by Saint Yvor*

THERE IS A DISTRICT called Ferneginan in Leinster. It is separated from Wexford by the river Slaney only. From there the larger mice that are commonly called rats were entirely expelled by the curse of the bishop, Saint Yvor, whose books they had happened to eat. They cannot be bred nor can they live there, if brought in.

66 ¶ *The fugitive bell*

THERE IS IN LEINSTER in the land of Mactalewus a certain bell, which, unless it is exorcized each night by its guardian with an exorcism especially composed for the purpose, and is bound by some chain, however light, is found in the morning at Clonard in Meath in the church of Saint Finian, whence it came. It is certain that this has happened on some occasions.

67 ¶ *Various miracles in Kildare; and first about the fire that never goes out and whose ashes do not increase*

IN KILDARE, in Leinster, which the glorious Brigid has made famous, there are many miracles worthy of being remembered. And the first of them that occurs to one is the fire of Brigid which, they say, is inextinguishable. It is not that it is strictly speaking inextinguishable, but that the nuns and holy women have so carefully and diligently kept and fed it with enough material, that through all the years from the time of the virgin saint until now it has never been extinguished. And although such an amount of wood over such a long time has been burned there, nevertheless the ashes have never increased.

68 ¶ *How Brigid keeps the fire on her own night*

ALTHOUGH in the time of Brigid there were twenty servants of the Lord here, Brigid herself being the twentieth, only nineteen have ever been here after her death until now, and the number has never increased. They all, however, take their turns, one each night, in guarding the fire. When the twentieth night comes, the nineteenth nun puts the logs beside the fire and says:

'Brigid, guard your fire. This is your night.'

And in this way the fire is left there, and in the morning the wood, as usual, has been burnt and the fire is still alight.

69 ¶ *The hedge around the fire that no male may cross*

THIS FIRE IS SURROUNDED by a hedge which is circular and made of withies, and which no male may cross. And if by chance one does dare to enter – and some rash people have at times tried it – he does not escape the divine vengeance. Only women are allowed to blow the fire, and then not with the breath of their mouths, but only with bellows or winnowing forks. Moreover, because of a curse of the saint, goats never have young here.

There are very fine plains hereabouts which are called 'Brigid's pastures', but no one has dared to put a plough into them. It is regarded as miraculous that these pastures, even though all the animals of the whole province have eaten the grass down to the ground, nevertheless when morning comes have just as much grass as ever. One might say that of these pastures was it written:

And all the day-long browsing of thy herds
Shall the cool dews of one brief night repair.[39]

70 ¶ *The falcon in Kildare that was tamed and domesticated*

FROM THE TIME of Brigid a noble falcon was accustomed to frequent the place, and to perch on the top of the tower of a church. Accordingly it was called by the people 'Brigid's bird', and was held in a certain respect by all.

This bird used to do the bidding of the townspeople or the soldiers of the castle, just as if it were tamed and trained in chasing, and, because of its own speed, forcing duck and other birds of the land and rivers of the plain of Kildare from the air down to the ground to the great delight of the onlookers. For what place was left to the poor little birds, when men held the land and the waters, and a hostile and terrible tyrant of a bird endangered the air?

A remarkable thing about this bird was that it did not allow any mate into the precincts of the church where it used to live. When the season of mating came, it went far away from its accustomed haunts and, finding a mate in the usual manner in the mountains near Glendalough, indulged its natural instincts there. When that was finished it returned alone to the church.

In this is showed a good example of honour to churchmen, especially when they are entrusted with divine office within the precincts of the church.

Exactly at the time of the first departure of Lord John from Ireland, that bird which had lived for so many generations, and had so agreeably added interest to the shrine of Brigid, having occupied itself without sufficient caution with the prey which it had caught, and having too little feared the approaches of men, was killed by a rustic with a staff which he had in his hand.

From which it is clear that one must ever fear in prosperity a turn of the tide, and that one should have little confidence in a daily life that is delightful and well loved.

71 ¶ *A book miraculously written*

AMONG ALL THE MIRACLES of Kildare nothing seems to me more miraculous than that wonderful book which they say was written at the dictation of an angel during the lifetime of the virgin.

This book contains the concordance of the four gospels according to Saint Jerome, with almost as many drawings as pages, and all of them in marvellous colours. Here you can look upon the face of the divine majesty drawn in a miraculous way; here too upon the mystical representations of the Evangelists, now having six, now four, and now two, wings. Here you will see the eagle; there the calf. Here the face of a man; there that of a lion. And there are almost innumerable other drawings. If you look at them carelessly and casually and not too closely, you may judge them to be mere daubs rather than careful compositions. You will see nothing subtle where everything is subtle. But if you take the trouble to look very closely, and penetrate with your eyes to the secrets of the artistry, you will notice such intricacies, so delicate and subtle, so close together and well-knitted, so involved and bound together, and so fresh still in their colourings that you will not hesitate to declare that all these things must have been the result of the work, not of men, but of angels.

72 ¶ *The composition of the book*

ON THE NIGHT before the day on which the scribe was to begin the book, an angel stood beside him in his sleep and showed him a drawing made on a tablet which he carried in his hand, and said to him:

'Do you think that you can make this drawing on the first page of the book that you are about to begin?'

The scribe, not feeling that he was capable of an art so subtle, and trusting little in his knowledge of something almost unknown and very unusual, replied:

'No.'

The angel said to him:

'Tomorrow tell your lady, so that she may pour forth prayers for you to the Lord, that he may open both your bodily and mental eyes so as to see the more keenly and understand the more subtly, and may direct your hands to draw correctly.'

All this was done, and on the following night the angel came again and held before him the same and many other drawings. By the help of the divine grace, the scribe, taking particular notice of them all, and faithfully committing them to his memory, was able to reproduce them exactly in the suitable places in the book. And so with the angel indicating the designs, Brigid praying, and the scribe imitating, that book was composed.

73 ¶ *The cross in Dublin that speaks and gives testimony to the truth*

AND NOW we shall set out those things that happened in more modern times.

There is in Dublin in the church of the Holy Trinity a cross of most wonderful power. It bears the figure of the Crucified. Not many years before the coming of the English, and during the time of the Ostmen, it opened its hallowed mouth and uttered some words. Many people heard it.

It had happened that a certain citizen had invoked it, and it alone, as the witness and surety of a certain contract. As time went on the other contracting party denied the agreement and completely and steadfastly refused to return the money which the other had given him according to the terms of the contract. The citizens, more in irony than for any serious reason, declared that they should go in a body to the aforesaid church and hear what the cross would say. The cross being adjured and called to witness, gave testimony to the truth.

74 ¶ *How the same cross became immovable*

WHEN EARL RICHARD[40] first came with an army to Dublin the citizens had forebodings of many evils, and fearing that their city would fall, and having no confidence in its defence, decided to make their escape by sea, and wanted to bring with them to the islands the cross of which we have been speaking. They tried every resource of effort and industry to give effect to this, but nevertheless the people of the whole city could not move it either by force or skill..

75 ¶ *How a penny offered to the cross twice jumped back and on the third occasion after confession remained; and of the iron greaves that were miraculously restored*

WHEN THE CITY was captured, an archer, among others, offered a penny to the cross, and as he turned to go away, was hit in the back by the penny flung after him immediately. He took it up and offered it to the cross a second time, but the same thing happened, while many people standing about looked on and wondered. Then the archer confessed before all that on that very day he had plundered the residence of the archbishop within the very precincts of that church. A penance was imposed upon him, and he returned whatever he had got from the archbishop's residence. He then brought back the same penny in great fear and awe for the third time to the cross. This time finally it remained and did not move.

A similar thing happened on the staff of Raymond,[41] the constable of earl Richard.

A young man of his household stole some iron greaves, and the whole household had declared on oath before the cross in the church of the Holy Trinity their innocence of the crime. Not long afterwards the young man, having returned from England where he had gone, threw himself, worn out and miserable, at the feet of Raymond. No one had suspected him of the crime. He offered satisfaction for the deed he had done and asked for pardon. He proclaimed openly and publicly that he had been so persecuted by the cross that ever since his perjury it had seemed to hang about his neck with a heavy weight, so that from then on he had not been able to sleep or have any rest.

With these and other manifestations and evidences of its power shown during the time of our arrival in the country, the cross has earned everywhere respect and veneration.

76 ¶ *The fanatic at Ferns who foretold the future from the past*

ABOUT THE TIME that the fitzMaurices got the castle at Ferns, a young man of their household, who had been called 'Fantasticus', plundered the church of Saint Maidoch, and being taken by a frenzy immediately went mad. By what spirit he was led, I do not know, but he began to prophesy and to foretell the future from the past.

'I can see,' he said, 'our men slain' (he mentioned several

of them by name) 'and all the castle laid low. It is no longer there.'

And so he continued to prophesy day after day, while everyone was astonished. In fact he did not cease until with the enemy's attack everything, in short, that he had predicted had come to pass as he had said.

77 ¶ *How an archer who crossed the hedge of Brigid went mad and how another lost his leg*

AT KILDARE an archer of the household of earl Richard crossed over the hedge and blew upon Brigid's fire. He jumped back immediately, and went mad. Whomsoever he met, he blew upon his face and said:

'See! That is how I blew on Brigid's fire.'

And so he ran through all the houses of the whole settlement, and wherever he saw a fire he blew upon it using the same words. Eventually he was caught and bound by his companions, but asked to be brought to the nearest water. As soon as he was brought there, his mouth was so parched that he drank so much that, while still in their hands, he burst in the middle and died.

Another who, upon crossing over to the fire, had put one shin over the hedge, was hauled back and restrained by his companions. Nevertheless the foot that had crossed perished immediately with its shank. Ever afterwards, while he lived, he was lame and feeble as a consequence.

78 ¶ *How seed wheat on being cursed by the bishop of Cork did not grow, and in the following year was miraculously interchanged with rye*

IN CORK a soldier took the land of Saint Finbar without the assent of the person in charge, ploughed it and scattered seed all over it. The bishop of the place came up and, invoking God and the saints of his church, prohibited him from forcibly occupying or sowing that land any more. The soldier nevertheless would not give up. The bishop then coming back to him said with tears:

'I ask almighty God that that seed may never come to harvest for you.'

And it happened that that year, to the astonishment of all the people of the city, not a single ear grew in those fields, nor any blade, nor grain. In the following year, however, other people, with the consent of the person in charge, sowed rye in the same place. When the harvest came they reaped pure wheat without any admixture of rye, either because the rye seed was miraculously changed to the nature of wheat, or rather because through the merits of the holy man the seed that did not produce the former year was preserved for the second year's crop.

79 ¶ *How Philip of Worcester was struck with sickness at Armagh, and how Hugh Tyrrell was scourged through divine intervention*

WHEN PHILIP OF WORCESTER was leading his army during Lent near Armagh, the seat of the blessed Patrick and the special see of the primacy of the whole of Ireland, and was forcibly taking a great tribute from the holy clergy during these days of prayer, he was returning with his spoils and gold, when suddenly within a short distance he was stricken with an illness and scarcely survived.

Hugh Tyrrell carried off to Louth with him a big cooking-pot that belonged to a community of clerics, and for doing so was cursed with the maledictions of all the clergy. But during that night a fire broke out in his lodgings, and two horses that had drawn the cooking-pot, and many other things, were burned immediately. Most of the settlement was also burned on that occasion. When Hugh Tyrrell saw this and found the cooking-pot in the morning undamaged, he repented and sent it back to Armagh. The bishop of Louth, then present, said of the same Hugh in the hearing of many of his army:

'That man will this year certainly meet with a grievous misfortune. Never was it known that the tears and curses of so many good men were spent in vain.'

And within the year we ourselves saw it happen to him because of a conflict that arose between himself and Hugo de Laci provoked by their own peoples. Almost the whole kingdom was entirely upset and overturned.

80 ¶ *The mill that will not grind on Sundays or anything stolen or plundered*

IN OSSORY there is a mill of Saint Luthernus, abbot, which does not grind on Sundays, nor does it grind anything got by theft or in plunder.

81 ¶ *The mill that women do not enter*

IN FORE in Meath there is a mill which Saint Fechin cut with his own hands out of the side of a rock. Women cannot enter this mill, no more than they can enter the same saint's church. The local people have as much reverence for this mill as they have for any of the churches of the saint.

It happened, however, when Hugo de Laci was leading his army through that place, that an archer dragged a woman into the mill and lustfully violated her there. He was stricken in his member with hell-fire in sudden vengeance and immediately began to burn throughout his whole body. He died the same night.

82 ¶ *How two horses on eating oats stolen from the same mill immediately died*

ONCE WHEN HIS ARMY was staying for the night in the same place, Hugo de Laci ordered his men to return the corn that they had stolen everywhere from the churches and from the mill. They restored all except a small quantity of oats which two soldiers left secretly in front of their horses. One of the horses went mad and died that night, having broken his head in the stable. And the other, while his rider was scoffing at the others who through a superstitious fear had returned the corn, fell dead on the following morning, suddenly and unexpectedly, beside Hugo de Laci. Most of the army saw it and were astounded.

83 ¶ *That the saints of this country seem to be of a vindictive cast of mind*

THIS SEEMS TO ME a thing to be noticed that just as the men of this country are during this mortal life more prone to anger and revenge than any other race, so in eternal death the saints of this land that have been elevated by their merits are more vindictive than the saints of any other region.

¶ Here begins the third part of the History of Ireland

84 ¶ *The Inhabitants of the country*[42]

FOR THE REST, IT NOW seems time to turn our pen to the description of the first inhabitants of this land, and take, one after the other, the arrivals of different peoples: how they came, and from where; how long they stayed in the island, and how they disappeared – we shall try to explain all this as briefly and as clearly as possible.

For a proper order seems to demand that after we have placed the land itself in the ocean, and outlined its position and nature, as well as its characteristics from the point of view of the presence or absence of various living things, we should then finally treat of man himself, as being the most worthy subject of our investigation, and on whose account we have treated of the other things. We shall give an account of his customs and manners here, and of the various happenings that occurred, according to the turns of fortune, down to our own times.

85 ¶ *The first arrival that namely of Cesara the grand-daughter of Noah before the Flood*

ACCORDING TO the most ancient histories of the Irish, Cesara, the grand-daughter of Noah, hearing that the Flood was about to take place, decided to flee in a boat with her companions to the farthest islands of the West, where no man had yet ever lived. She hoped that the vengeance of the Flood would not reach to a place where sin had never been committed. All the ships of her company were wrecked. Her ship alone, carrying three men and fifty women, survived. It put in at the Irish coast, by chance, one year before the Flood.

All the same, in spite of her cleverness, and, for a woman, commendable astuteness in seeking to avoid evil, she did not succeed in putting off the general, not to say universal, disaster.

The place where that ship first put in is called 'the shore of the ships', and the place where the aforementioned Cesara is buried is called even now 'Cesara's tomb'.

But since almost all things were destroyed in the Flood, one may reasonably have doubts as to the value of the account of these arrivals and events that has been handed down after the Flood. Let those, however, who first wrote these accounts be responsible. My function is to outline, not to attack, such stories. Perhaps, as in the case of the invention of music before the Flood, a narrative of these things had been inscribed on some material, stone or tile, which later was found and preserved.

86 ¶ *The second arrival that namely of Bartholanus (Parthalón) three hundred years after the Flood*

IN THE THREE HUNDREDTH YEAR after the Flood Bartholanus, the son of Sera, of the stock of Japhet, the son of Noah, is said to have been brought to the coasts of Ireland with his three sons and their wives, either by chance or through their own efforts – that is to say, either through losing their way, or having had a good idea of their future fatherland. His three sons were called Langninus, Salanus, and Ruturugus. These three, in leaving traces of their names on certain things still existing, have aquired an immortal memory, and, so to speak, a continual and living presence.

After the first is named the lake of Lagilinus,[43] and after the second is named the high mountain that overlooks the sea between Ireland and Britain – Salanga. Now, however, this latter is more usually called Dominic's mountain,[44] because Saint Dominic built a fine monastery at its foot at a much later time. Ruturugus, the third in the succession, gave his name to lake Ruturugus.[45]

We find few outstanding things done during the time of this Bartholanus, and indeed almost nothing, except that four enormous lakes suddenly burst forth from the bowels of the earth; and, in the interests of developing agriculture, four huge forests were cut away from the very roots, and so an open plain was made – but not without the sweat of many. For at that time almost the whole land, with the exception of a few scattered mountain districts, was overgrown with vast and ever-multiplying woods. Consequently there was scarcely any open place where one could ply the plough. Even to this day the plains here are few in proportion to the woods.

Bartholanus, however, and his sons and descendants were fortunate not only in their material successes but also in the increase of their stock. For after a cycle of three hundred years from their arrival the number of their descendants is said to have increased to nine thousand. At length, having conquered the giants in a great war, and because human prosperity cannot last for ever, and because 'the gods easily grant great blessings, but do not easily maintain them',[46] and 'the gods have imposed this limit to increasing success', and 'the great come into conflict with one another', and 'to persist for long is denied to the great: heavy indeed is the fall under a great weight',[47] he died with nearly all his descendants as a result of a sudden pestilence and, perhaps, the corruption of the air arising from the corpses of the giants that had been killed. Only Ruanus is said to have escaped the disaster.

This Ruanus, as the ancient histories recount, survived more cycles of years than can easily be believed right down to the time of Saint Patrick by whom he had been baptized. He is said to have given to Saint Patrick a full account of the history of Ireland and all the deeds of his people, that had almost, because of their great antiquity, escaped the folds of memory. For there is nothing so fixed in the mind but neglect and the passage of time will make to fade away. This Ruanus, although he succeeded in making an armistice with death, nevertheless did not succeed in making a peace. He pushed forward the barrier of his life far beyond the ordinary and accustomed limit of this fleeting and short existence. Nevertheless he did not escape the fate that falls on mortal flesh.

Ruanus exceeded the ordinary term of life.

87 ¶ *The third arrival that namely of Nemedus from Scythia with his four sons*

THE AFORESAID BARTHOLANUS and all his race having been wiped out by the sword of a cruel and long-enduring pestilence, the land remained for some time deprived of anyone to inhabit it, until Nemedus, the son of Agnominius a Scythian, arrived with his four sons at the shore of the abandoned land. His sons were Starvus, Gerbaneles, Anninus, and Fergusius.[48]

In the time of Nemedus four lakes suddenly and with irresistible force emerged, and many woods and groves were changed to fields and plains. He also waged four fierce wars with pirates that were accustomed to plunder the country. He was victorious everywhere. He died, however, in an island in the south of Ireland, to which he left his name as a lasting heritage.

His sons and grandsons and great-grandsons multiplied in a short time to such an extent that they filled every corner of the whole island with more inhabitants than it ever had, but since 'what grows gradually is more solid', and 'one cannot usually rely on what has sprung up suddenly',[49] so, just as they had increased to such a great number, they were destroyed much more quickly than they had arisen through sudden and unexpected happenings. For most of them were quickly destroyed in the frequent wars that they waged with the giants who then flourished in the island, and through various diseases and plagues. Those that were left, seeking refuge in flight from the many evils that threatened them, made in their boats some for Scythia, and some for Greece.

The race of Nemedus held Ireland for two hundred and sixteen years, and for two hundred years afterwards the land was empty.

88 ¶ *The fourth arrival that namely of the five brothers and sons of Dela*[50]

WHEN THESE THINGS had happened in this order, at length five chieftains who were at the same time brothers, the sons of Dela, springing from the branch of the people of Nemedus that had taken refuge in Greece, landed in Ireland. They found it uninhabited and divided it in five equal portions among themselves.

The bounds of these divisions meet at a certain stone in Meath near the castle of Kilair (Killare).[51] This stone is said to be the navel of Ireland, as it were, placed right in the middle of the land. Consequently that part of Ireland is called Meath, as being situated in the middle of the island. But Meath does not belong to any one of the five well-known divisions of which we have spoken above.

These five brothers were the first to divide the island into five portions.

When the aforesaid five brothers, namely Gandius, Genandius, Sangandius, Rutheraigus, and Slanius[52] had divided the island into five portions, each portion had a little section in Meath touching upon the stone. The reason was that, as that land was from the beginning best, both by reason of its being a plain and on account of its rich harvests, none of them wished to be left without a share in it.

89 ¶ *The first king of Ireland, Slanius*

AS TIME WENT ON, and fortune varied and, as is her wont, turned many things upside down in a short time, Slanius became the sole king of the whole of Ireland. As a result he is called the first king of Ireland. He first put the five small portions of Meath together, treated it as a unit, and assigned it to the royal demesne. Consequently Meath even to this day remains separate from those other five principal portions which, as has been said, were destroyed by Slanius. It does not contain as much land as any of the other five but only half as much.

From the time of Slanius each of the five portions contains thirty-two cantreds, but Meath contains only sixteen. There are, therefore, eighty [*sic*] plus sixteen cantreds in all. Cantred is a word used in the normal way in the Irish and British tongues for as much land as usually holds a hundred settlements.

Since from among these brothers and their descendants nine kings followed, their reign was short and lasted only about thirty years. Slanius was buried in a hill in Meath which takes its name from him.

90 ¶ *The fifth arrival that namely of the four sons of king Milesius from Spain, and how Herimon and Heberus divided the kingdom between them*

THESE PEOPLE having been for the most part destroyed and very much weakened, not only because of internal conflicts, but especially on account of a war which they waged with other descendants of Nemedus that had come from Scythia, and from which they had greatly suffered, eventually four noble sons of king Milesius came from

Spain in a fleet of sixty ships. They immediately conquered the whole island without any opposition. As time went on the two outstanding among them, Heberus and Herimon, divided the kingdom into two equal parts between themselves. Herimon got the southern, and Heberus the northern part.

91 ¶ *The discord between the two brothers and how Heberus having been killed Herimon was the first king of the Irish people*

WHEN THEY HAD REIGNED for some time prosperously and happily together, since 'sharers in a kingdom are not to be relied on, and all power is impatient of being shared',[53] blind ambition, the mother of evils, gradually undermined the agreement of the brothers and quickly dissolved the bond of peace. Discord raised her head where all had been well and succeeded in upsetting and destroying everything.

After there had been a number of conflicts and many doubtful issues in wars, Herimon eventually won. His brother Heberus was killed in war and so the whole kingdom fell to him alone. He was the first king of the Irish people that to this day inhabit the island.

The Hibernienses, according to some, take their name from the Heberus of whom we have been speaking. But others with greater probability say that they take their name from the Spanish river Hiberus whence they came. They are also called Gaideli and Scoti. As the old histories tell us, a certain Gaidelus, a descendant of Phenius,[54] having, after the episode of the confusion of tongues at the tower of Nembrotica, become very skilled in various languages, was, on account of that skill, joined in marriage by king Pharao to his daughter Scotia. Since then the Hibernienses derive, as they say, their line from these two, Gaidelus and Scotia, they

From this Heberus or rather the Spanish river Hiberus the Hibernienses took their name.

The Hibernienses are called Gaideli from Gaidelus, the grandson of Phenius, and Scoti from Scotia, his wife.

are called, as they are born, Gaideli and Scoti. This Gaidelus, they say, composed the Irish language, which is called Gaidelach, as much as to say that it was brought together from all languages.

The northern part of Britain is also called Scotia, because it is known to be inhabited by a people which was originally propagated by Gaidelus and Scotia. The affinity in language and culture, as well as in weapons and customs, to this day bears out this fact.

The northern part of Britain is called Scotia, because it was inhabited by a people of the Scoti.

92 ¶ *Gurguintius king of the Britons who sent the Basclenses*[55] *to Ireland and gave it to them to settle in*

AS THE BRITISH HISTORY[56] relates, the king of the Britons, Gurguintius, son of the noble Belinus and grandson of the famous Brennius, when returning from Denmark, which his father had formerly conquered, and which, when it had rebelled, he himself had again brought into subjection, found at the Orkney Islands a fleet which had brought Basclenses there from Spain. Their leaders approached the king, and told him whence they had come and the reason for their coming, namely to settle in a country of the West. They were urgent in their request that he should give them some land to inhabit. Eventually the king, on the advice of his counsellors, gave them that island that is now called Ireland, and which was then either entirely uninhabited or had been settled by him. He also gave them pilots for their expedition from among his own fleet.

From this it is clear that Ireland can with some right be claimed by the kings of Britain, even though the claim be from olden times.

Note: The twofold claim from olden times.

The kings of Britain have a right to Ireland.

Secondly, the city of Bayonne is on the boundary of Gascony, and belongs to it. It is also the capital of Basclonia, whence the Hibernienses came. And now Gascony and all Aquitaine rejoices in the same rule as Britain.[57]

The twofold claim from recent times

The kings of Britain have also a newly established double claim. On the one hand the spontaneous surrender and protestation of fealty of the Irish chiefs – for everyone is allowed to renounce his right; and on the other, the favour of the confirmation of the claim by the Pope.[58]

For when Jupiter started thundering in the confines of the western ocean, the petty Western kings were frightened by the thunder and averted the stroke of the thunderbolt by sheltering from it in a peace. But about these matters more will be said in the proper place.

93 ¶ *The nature, customs and characteristics of the people*

I HAVE THOUGHT IT not superfluous to say a few things about the nature of this people both in mind and body, that is to say, of their mental and physical characteristics.

To begin with: when they are born, they are not carefully nursed as is usual. For apart from the nourishment with which they are sustained by their hard parents from dying altogether, they are for the most part abandoned to nature. They are not put in cradles, or swathed; nor are their tender limbs helped by frequent baths or formed by any useful art. The midwives do not use hot water to raise the nose, or press down the face, or lengthen the legs. Unaided nature according to her own judgement arranges and disposes without the help of any art the limbs that she has produced.

As if to prove what she can do by herself she continually shapes and moulds, until she finally forms and finishes them in their full strength with beautiful upright bodies and handsome and well-complexioned faces.

But although they are fully endowed with natural gifts, their external characteristics of beard and dress, and internal cultivation of the mind, are so barbarous that they cannot be said to have any culture.

The Hibernienses by nature's gift are handsome, but shameful in their practices and culture.

They use very little wool in their dress and that itself nearly always black – because the sheep of that country are black – and made up in a barbarous fashion. For they wear little hoods, close-fitting and stretched across the shoulders and down to a length of about eighteen to twenty-two inches, and generally sewn together from cloths of various kinds. Under these they wear mantles instead of cloaks. They also use woollen trousers that are at the same time boots, or boots that are at the same time trousers, and these are for the most part dyed.

When they are riding, they do not use saddles or leggings or spurs. They drive on, and guide their horses by means of a stick with a crook at its upper end, which they hold in their hand. They use reins to serve the purpose both of a bridle and a bit. These do not keep the horses, accustomed to feeding on the grass, from their food.

Moreover, they go naked and unarmed into battle. They regard weapons as a burden, and they think it brave and honourable to fight unarmed. They use, however, three types of weapons – short spears, two darts (in this they imitate the Basclenses), and big axes well and carefully forged, which they have taken over from the Norwegians and the Ostmen, about which we shall speak later.

Three types of weapons.

They are quicker and more expert than any other people in throwing, when everything else fails, stones as missiles, and such stones do great damage to the enemy in an engagement.

They are a wild and inhospitable people. They live on beasts only, and live like beasts. They have not progressed at all from the primitive habits of pastoral living.

While man usually progresses from the woods to the

fields, and from the fields to settlements and communities of citizens, this people despises work on the land, has little use for the money-making of towns, contemns the rights and privileges of citizenship, and desires neither to abandon, nor lose respect for, the life which it has been accustomed to lead in the woods and countryside.

They use the fields generally as pasture, but pasture in poor condition. Little is cultivated, and even less sown. The fields cultivated are so few because of the neglect of those who should cultivate them. But many of them are naturally very fertile and productive. The wealth of the soil is lost, not through the fault of the soil, but because there are no farmers to cultivate even the best land: 'the fields demand, but there are no hands'.[59] How few kinds of fruit-bearing trees are grown here! The nature of the soil is not to be blamed, but rather the want of industry on the part of the cultivator. He is too lazy to plant the foreign types of trees that would grow very well here.

Because of laziness no progress from pastoral living has been made; fields are not cultivated; there is no produce from varieties of fruit-bearing trees and mines.

The different types of minerals too, with which the hidden veins of the earth are full, are not mined or put to any use, precisely because of the same laziness. Even gold, of which they are very desirous – just like the Spaniards – and which they would like to have in abundance, is brought here by traders that search the ocean for gain.

They do not devote their lives to the processing of flax or wool, or to any kind of merchandise or mechanical art. For given only to leisure, and devoted only to laziness, they think that the greatest pleasure is not to work, and the greatest wealth is to enjoy liberty.

This people is, then, a barbarous people, literally barbarous. Judged according to modern ideas, they are uncultivated, not only in the external appearance of their dress, but also in their flowing hair and beards. All their habits are the habits of barbarians. Since conventions are formed from living together in society, and since they are so removed in these distant parts from the ordinary world of men, as if they were in another world altogether and consequently cut off from well-behaved and law-abiding people, they know only of the barbarous habits in which they were born and brought up, and embrace them as another nature. Their natural qualities are excellent. But almost everything acquired is deplorable.

94 ¶ *The incomparable skill of the people in musical instruments*

IT IS ONLY in the case of musical instruments that I find any commendable diligence in the people. They seem to me to be incomparably more skilled in these than any other people that I have seen.

The movement is not, as in the British instrument to which we are accustomed, slow and easy, but rather quick and lively, while at the same time the melody is sweet and pleasant. It is remarkable how, in spite of the great speed of the fingers, the musical proportion is maintained. The melody is kept perfect and full with unimpaired art through everything – through quivering measures and the involved use of several instruments – with a rapidity that charms, a rhythmic pattern that is varied, and a concord achieved through elements discordant. They harmonize at intervals of the octave and the fifth, but they always begin with B flat and with B flat end, so that everything may be rounded with the sweetness of charming sonority. They glide so subtly from one mode to another, and the grace notes so freely sport with

such abandon and bewitching charm around the steady tone of the heavier sound, that the perfection of their art seems to lie in their concealing it, as if 'it were the better for being hidden. An art revealed brings shame.'[60]

Hence it happens that the very things that afford unspeakable delight to the minds of those who have a fine perception and can penetrate carefully to the secrets of the art, bore, rather than delight, those who have no such perception – who look without seeing, and hear without being able to understand. When the audience is unsympathetic they succeed only in causing boredom with what appears to be but confused and disordered noise.

The instruments of Wales and Scotland.

One should note that both Scotland and Wales, the former because of her affinity and intercourse, the latter as it were by grafting, try to imitate Ireland in music and strive in emulation. Ireland uses and delights in two instruments only, the harp, namely, and the timpanum.[61] Scotland uses three, the harp, timpanum, and the crowd. Wales uses the harp, the pipes, and the crowd. They also use strings made of bronze, and not from leather. In the opinion, however, of many, Scotland has by now not only caught up on Ireland, her instructor, but already far outdistances and excels her in musical skill. Therefore people now look to that country as to the fountain of the art.

95 ¶ *How many kings reigned from Herimon to the coming of Patrick and how the island was converted to the Faith by him*

FROM THE TIME of the first king of this people, namely Herimon, until the coming of Patrick, one hundred and thirty-one kings of the same people reigned in Ireland.

Patrick, Britannic by birth,[62] a man distinguished for his life and holiness, arrived in the island, and, finding the people given to idolatry and deluded by various errors, was the first by the aid of divine grace, to preach and plant there the Christian Faith. He baptized the people, whole crowds at a time, and, the entire island having been converted to the Faith of Christ, chose Armagh as his see. He made this place a kind of metropolis and special seat for the primacy of the whole of Ireland. He also appointed bishops in suitable places, so that, having been called to share his responsibility, they should water what he had planted. The Lord, however, would grant the increase.

96 ¶ *How there were no archbishops in Ireland before the arrival of John Papiro who established four* pallia *in Ireland*

THERE WERE NO ARCHBISHOPS in Ireland,[63] but the bishops just consecrated one another, until John Papiro, legate of the Roman see, came here not many years ago.[64] He established four *pallia* in Ireland. One he gave to Armagh; another to Dublin; a third to Cashel; and the fourth to Tuam in Connacht.

The blessed Patrick died and slept in the Lord in the one hundred and twentieth year of his age, four hundred and fifty-eight years after the incarnation of the Lord, and one thousand eight hundred years after the coming of the Hibernienses.

97 ¶ *How the three bodies of Patrick, Columba and Brigid were in our own times found in Ulster in the city of Down and were translated*

SAINT COLUMBA and Saint Brigid were contemporaries of Patrick.[65] Their three bodies were buried in Ulster in the same city, namely, Down. They were found there in our times, in the year, that is, that Lord John first came to Ireland, in a cave that had three sections. Patrick was lying in the middle, and the others were lying one on either side. John de Courci, who was in command there, took charge when these three noble treasures were, through divine revelation, found and translated.

98 ¶ *The Irish are ignorant of the rudiments of the Faith*

ALTHOUGH SINCE THE TIME of Patrick and through so many years the Faith has been founded in the island, and has almost continuously thrived, it is, nevertheless, remarkable that this people even still remains so uninstructed in its rudiments.

This is a filthy people, wallowing in vice. Of all peoples it is the least instructed in the rudiments of the Faith. They do not yet pay tithes or first fruits or contract marriages. They do not avoid incest. They do not attend God's church with due reverence. Moreover, and this is surely a detestable thing, and contrary not only to the Faith but to any feeling of honour – men in many places in Ireland, I shall not say marry, but rather debauch, the wives of their dead brothers. They abuse them in having such evil and incestuous relations with them. In this (wishing to imitate the ancients more eagerly in vice than in virtue) they follow the apparent teaching, and not the true doctrine, of the Old Testament.

99 ¶ *Their vices and treacheries*

MOREOVER, above all other peoples they always practise treachery. When they give their word to anyone, they do not keep it. They do not blush or fear to violate every day the bond of their pledge and oath given to others – although they are very keen that it should be observed with regard to themselves. When you have employed every safeguard and used every precaution for your own safety and security, both by means of oaths and hostages, and friendships firmly cemented, and all kinds of benefits conferred, then you must be especially on your guard, because then especially their malice seeks a chance. For they feel that because of your reliance on your safeguards you are not on the watch. Then at last they resort to the arts of evil and their accustomed weapons of deceit, so that, taking an opportunity of your feeling of security, they may be able to injure you when you do not expect it.

You must be more afraid of their wile than their war; their friendship than their fire; their honey than their hemlock; their shrewdness than their soldiery; their betrayals than their battle lines; their specious friendship than their enmity despised. For this is their principle: 'Who asks of an enemy whether he employs guile or virtue?'[66] These are their characteristics: they are neither strong in war, nor reliable in peace.

100 ¶ *How they always carry an axe as if it were a staff in their hand*

FROM AN OLD and evil custom they always carry an axe in their hand as if it were a staff. In this way, if they have a feeling for any evil, they can the more quickly give it effect. Wherever they go they drag this along with them. When they see the opportunity, and the occasion presents itself, this weapon has not to be unsheathed as a sword, or bent as a bow, or poised as a spear. Without further preparation, beyond being raised a little, it inflicts a mortal blow. At hand, or rather, in the hand and ever ready is that which is enough to cause death.

From the axe there is always anxiety. If you think that you are free from anxiety, you are not free from an axe. You admit a risk, if you admit an axe, and are free from anxiety. If they see an opportunity of exercising their evil, it would have been better that they had not seen it, or rather that they had not seen anything at all.

101 ¶ *A proof of their wickedness and a new way of making a treaty*

AMONG MANY OTHER TRICKS devised in their guile, there is this one which serves as a particularly good proof of their treachery.

Under the guise of religion and peace they assemble at some holy place with him whom they wish to kill. First they make a treaty on the basis of their common fathers. Then in turn they go around the church three times. They enter the church and, swearing a great variety of oaths before relics of saints placed on the altar, at last with the celebration of Mass and the prayers of the priests they make an indissoluble treaty as if it were a kind of betrothal. For the greater confirmation of their friendship and completion of their settlement, each in conclusion drinks the blood of the other which has willingly been drawn especially for the purpose.

O! how often in the very hour of this alliance has blood been so treacherously and shamefully shed by treacherous blood relations that one or other has been left entirely drained of blood. O! how often a bloody divorce immediately follows within the same hour, or precedes, or even – and this is unheard of elsewhere – interrupts the very ceremony of the 'betrothal'!

Woe to brothers amongst a barbarous people! Woe to kinsmen! When they are alive they are relentlessly driven to death. When they are dead and gone, vengeance is demanded for them. If this people has any love or loyalty it is kept only for foster children and foster brothers.

They love foster children and foster brothers and drive kinsmen to death.

To such an extent does one seem here to be allowed to carry out whatever one desires; people are so concerned not with what is honourable, but all of them only with what is expedient (although in fact only what is honourable can be said to be entirely expedient); so strongly has the pest of treachery grown and put in roots here; so natural through long usage have bad habits become; to such an extent are habits influenced by one's associates, and he who touches pitch will be defiled by it; that foreigners coming to this country almost inevitably are contaminated by this, as it were, inborn vice of the country – a vice that is most contagious.

Foreigners are contaminated by the same vice.

This place finds people already accursed or makes them so. For since the road to pleasure is downhill, and nature tends to imitate vice, who has any hesitation about going on the road to perdition, when he is persuaded and convinced by so many examples of sacrilegious men, so many evidences of evil deeds, such frequent transgression of oaths, such complete lack of respect for the Faith, and is continually being invited to do similar things by a precept that inculcates evil?

102 ¶ *A new and outlandish way of confirming kingship and dominion* [67]

THERE ARE SOME THINGS which, if the exigencies of my account did not demand it, shame would

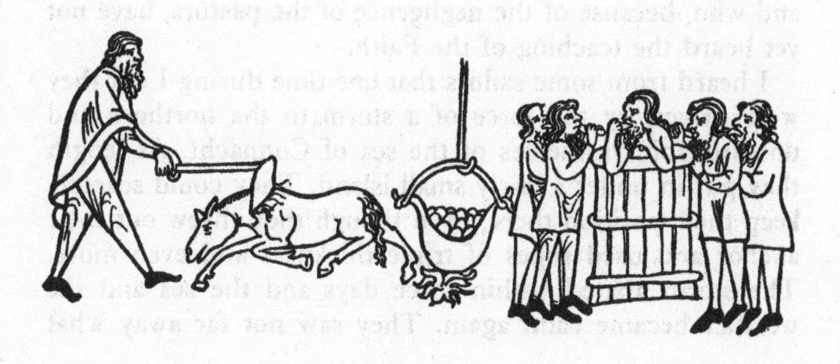

discountenance their being described. But the austere discipline of history spares neither truth nor modesty.

There is in the northern and farther part of Ulster, namely in Kenelcunill, a certain people which is accustomed to appoint its king with a rite altogether outlandish and abominable. When the whole people of that land has been gathered together in one place, a white mare is brought forward into the middle of the assembly. He who is to be inaugurated, not as a chief, but as a beast, not as a king, but as an outlaw, has bestial intercourse with her before all, professing himself to be a beast also. The mare is then killed immediately, cut up in pieces, and boiled in water. A bath is prepared for the man afterwards in the same water. He sits in the bath surrounded by all his people, and all, he and they, eat of the meat of the mare which is brought to them. He quaffs and drinks of the broth in which he is bathed, not in any cup, or using his hand, but just dipping his mouth into it round about him. When this unrighteous rite has been carried out, his kingship and dominion have been conferred.

103 ¶ *Many in the island have never been baptized, and have not yet heard of the teaching of the Faith*

MOREOVER, although all this time the Faith has grown up, so to speak, in the country, nevertheless in some corners of it there are many even still who are not baptized, and who, because of the negligence of the pastors, have not yet heard the teaching of the Faith.

I heard from some sailors that one time during Lent they were driven by the force of a storm to the northern and unsearchable vastnesses of the sea of Connacht. At length they put in under a fairly small island. They could scarcely keep their position there, even though they threw out their anchor and used ropes of triple thickness and even more. The storm abated within three days and the sea and the weather became calm again. They saw not far away what appeared to them to be land that was completely unknown to them.

Shortly afterwards they caught sight of a small skiff putting out from the land towards themselves. The boat was narrow and oblong, made of wickerwork and covered on the outside with sewn hides of animals. There were two men in the boat who were altogether naked except for broad belts of the raw hides of animals which they had tied about their waists. Their hair was very long and flaxen, coming down and across their shoulders, as is the Irish manner, and covering most of their bodies. When they had found out from them that they were from some part of Connacht, and spoke the Irish language, they took them on board the ship. They, on their part, began to wonder at everything they saw as if it were new.

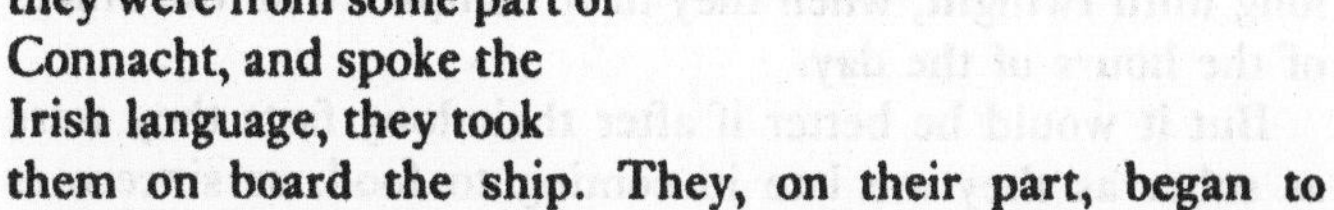

They said that they had never before seen a big ship made of wood, nor the trappings of civilization. When they were offered bread and cheese to eat they did not know what they were and refused them. They said that they fed only on meat, fish, and milk. They did not usually wear clothes, but sometimes in great necessity they used the hides of animals. They asked the sailors if they had any meat for a meal on board, and when they were told that it was not allowed to eat meat during Lent, they knew nothing about Lent. Nor did they know anything about the year, nor the month, nor the week, and they were completely ignorant of the names of the days of the week. When they were asked if they were Christians and baptized, they replied that they had as yet heard nothing of Christ and knew nothing about him.

And so they departed and took with them one piece of bread and a cheese, so that they might show to their own people as a wonder the kind of food that other peoples used.

104 ¶ *The Irish clergy, in many points praiseworthy*[68]

NOW WE SHALL TURN our attention to the clergy. The clergy of this country are on the whole to be commended for their observance. Among their other virtues chastity shines out as a kind of special prerogative. They diligently carry out their obligations in the matter of the Psalms and the hours, reading and praying. They keep themselves within the enclosures of the church and fulfil the divine offices with which they are entrusted. They practise a considerable amount of abstinence and asceticism in the use of food. Most of them, in fact, fast daily all day long until twilight, when they have completed all the offices of the hours of the day.

But it would be better if after their long fasts they were as sober as they are late in coming to food, as sincere as they are severe, as pure as they are dour, and as genuine as they appear.

105 ¶ *The prelates should be reproved for their neglect of their pastoral office*

I HAVE almost only one thing on which to reprove the bishops and prelates, and that is that they are too slack and negligent in the correction of a people that is guilty of such enormities. Because of their not preaching to and reproving their people, I preach that they should be reproved themselves. For the fact that they do not charge others with evil deeds, I charge them. For not reprehending others I reprehend them. If the prelates from the time of Patrick through all those years had done a man's job, as they should have done, in preaching and instructing, chastising and correcting, they would have extirpated at any rate to a certain extent those abominations of the people already mentioned, and would have impressed upon them some semblance of honour and religious feeling.

But there was no one among them to raise his voice as a trumpet. There was none to mount on the other side, and be a wall for the house of Israel. There was none to fight for the church of Christ even to exile, to say nothing of blood – that church which Christ had purchased for himself with his precious blood.

Consequently all the saints of this country are confessors, and there is no martyr. It would be difficult to find such a state of things in any other Christian kingdom. There was found no one in those parts to cement the foundations of the growing church with the shedding of his blood. There was no one to do this service; not a single one.

All the saints of Ireland are confessors and there is no martyr.

For they are pastors that wish to be fed, and do not wish to feed. They are prelates that do not wish to be of use, but rather to use. And they are bishops who welcome the honour and name of their calling, but do not welcome its duties and responsibilities.

The prelates of this land, keeping themselves according to an old custom within the enclosures of their churches, give themselves almost always to contemplation alone. They are so enamoured of the beauty of Rachel that they find the blear-eyed Leah disgusting. Whence it happens that they neither preach the word of the Lord to the people, nor tell them of their sins, nor extirpate vices from the flock committed to them, nor instil virtues.

Since nearly all the prelates of Ireland are taken from monasteries into the ranks of the clergy, they scrupulously fulfil all the obligations of a monk. But they omit almost everything to which they are obliged as clerics and prelates. They care for and are mindful of themselves only, but they omit or put off with great negligence the care of the flock committed to them. They either do not know – or conveniently forget – the famous remark of Jerome to the monk Rusticus:[69] 'So live in a monastery that you may deserve to be a cleric. Learn over a long time that which later you will teach. And of those who are good, always follow those that are better. And when you have been made a cleric, do those things that a cleric should do'; and also the other advice to the same man: 'If the desire to be a cleric entices you, first learn what later you may be able to teach. Do not try to be a soldier before being a recruit, or a master before being a pupil.' But, indeed, these prelates make poor provision for, and take bad care of, themselves, when they carelessly and negligently deprive and rob of care and provision those whose care they have assumed in the office which they have undertaken. They lead themselves much more seriously and fatally astray than others.

106 ¶ *How the clergy differ from monks and are to be placed above them*

THEY OUGHT TO KNOW that as Jerome says to Eleutherius,[70] 'the obligation of a monk is one thing, that of a cleric is another. Clerics feed sheep, but monks are fed.' Monks are with respect to clerics what a flock is with respect to its shepherds. A monk is so called as being the guardian of one single individual, and looks after himself alone. But a cleric is obliged to assume responsibility for the care of many. The monk is like a grain of wheat remaining alone. But the cleric is like a grain that germinates and brings a big yield into the granary of the Lord.

Prelates of this kind, having two allegiances, have some obligations deriving from their being monks, and others deriving from their being clerics. As monks they should possess a dove-like simplicity; as clerics, the shrewdness of the serpent; on the one hand wisdom, on the other eloquence; on the one hand, words, on the other, deeds; on the one hand, a conscience, on the other, knowledge. And so let them fulfil both sets of obligations that, entering the tabernacle with the priests, their vestments may be hung about with little bells, and the words of instruction and reproof may be heard sounding from their lips.

Jerome clearly rebukes simple and silent prelates that have more of the monk than the cleric in them, when he says:[71] 'An innocent life without much speaking does as much damage by its silence as good by its example. For the wolves have to be driven off by the barking of the dogs and the staffs of the shepherds.' And he says much the same in his first Prologue to the Bible:[72] 'A holy simplicity does good only to itself; and it does as much damage to the church of God in not resisting those who seek to destroy it, as good through the merits of its life.' 'For the error that is not opposed is approved, and when truth is not defended, it is overwhelmed.'[73] 'To omit to harass those who are in error, when one can, is nothing else than to shield them. And one leaves one's self open to the charge of a secret understanding with the perpetrators, when one ceases to fight against an open evil'[74] – especially when one is bound to do so by one's office.

107 ¶ *A sly reply of the archbishop of Cashel*

WHEN ONCE UPON A TIME I was making these complaints and others like them to Tatheus, the archbishop of Cashel, a learned and discreet man, in the presence of Gerard, a cleric of the church of Rome, who was then on some mission or other in those parts, and was blaming the prelates especially for the terrible enormities of the country, using the very strong argument that no one had ever in that kingdom won the crown of martyrdom in defence of the church of God, the archbishop gave a reply which cleverly got home – although it did not rebut my point: 'It is true,' he said, 'that although our people are very barbarous, uncivilized, and savage, nevertheless they have always paid great honour and reverence to churchmen, and they have never put out their hands against the saints of God. But now a people has come to the kingdom which knows how, and is accustomed, to make martyrs. From now on Ireland will have its martyrs, just as other countries.'

From now on Ireland will have its martyrs, just as other countries.

108 ¶ *About bells, croziers and other such relics of the saints regarded by the people of Wales as well as by the people of Ireland with great reverence, and about a priest stricken with a double sickness*

I SHOULD NOT OMIT to mention also that the people and clergy of both Wales and Ireland have a great reverence for bells that can be carried about, and staffs belonging to the saints, and made of gold and silver, or bronze, and curved at their upper ends. So much so that they fear to swear or perjure themselves in making oaths on these, much more than they do in swearing on the gospels. For through some hidden power, as it were divinely given to these, and vindictiveness, in which the saints of this country seem to be very interested, those who show disrespect for these objects are usually chastised, and those that have transgressed are severely punished.

I saw a poor Irish mendicant in Wales – which makes it more remarkable – wearing on his neck as a relic a horn made of bronze which, he said, had belonged to Saint Brendan. He said that no one dared to sound it because of reverence for the saint. When, as was the custom in Ireland, he held it out to the people standing about to be kissed, a priest named Bernard snatched it from his hand. He put it to his mouth and started to blow, and make it sound. He was struck within the hour in the presence of many with a double sickness. The wretch had had a torrent of eloquence before then, and the loose tongue of a tale-bearer, but immediately he lost all use of speech. As a result he was so harmed that he has always had a speech impediment. In addition he went into a coma and immediately lost his memory altogether. He could scarcely remember that he had a name. He was in fact so injured in his memory that, as I myself saw, for many days afterwards he tried to memorize, as if anew, the Psalms, which before that he had off by heart perfectly. And I marvelled to see one who had had a sufficiently wide acquaintance with letters, as an old man at the end of his days trying to pick up what help he could get, just like one at the elementary stage.

109 ¶ *The great numbers among this people that are maimed in body*

MOREOVER, I have never seen among any other people so many blind by birth, so many lame, so many maimed in body, and so many suffering from some natural defect. Just as those that are well formed are magnificent and second to none, so those that are badly formed have not their like elsewhere. And just as those who are kindly fashioned by nature turn out fine, so those that are without nature's blessing turn out in a horrible way.

And it is not surprising if nature sometimes produces such beings contrary to her ordinary laws when dealing with a people that is adulterous, incestuous, unlawfully conceived and born, outside the law, and shamefully abusing nature herself in spiteful and horrible practices. It seems a just punishment from God that those who do not look to him with the interior light of the mind, should often grieve in being deprived of the gift of the light that is bodily and external.

110 ¶ *How many kings reigned from the time of Patrick to the coming of Turgesius*

FROM THE COMING of Patrick, therefore, until the times of king Fedlimidius (Feidhlimidh)[75] thirty-three kings of that people reigned for four hundred years in Ireland. During their time the Christian Faith remained here unimpaired and unshaken.

111 ¶ *How during the reign of Fedlimidius the Norwegians under the leadership of Turgesius conquered Ireland*

BUT IN THE REIGN of this Fedlimidius the Norwegians put in at the Irish shores with a great fleet. They both took the country in a strong grip and, maddened in their pagan fury, destroyed nearly all the churches. Their leader, who was called Turgesius, quickly subjected the whole island to himself in many varied conflicts and fierce wars. He journeyed throughout the whole country and strengthened it with strong forts in suitable places.

And so to this day, as remains and traces of ancient times, you will find here many trenches, very high and round and often in groups of three, one outside of the other, as well as walled forts which are still standing, although now empty and abandoned. For the people of Ireland have no use for castles. Woods are their forts and swamps are their trenches.

Turgesius ruled the kingdom of Ireland for some time in peace, until he died deceived by a trick involving girls.

112 ¶ *How the English say that it was Gurmundus, the Irish Turgesius, that subjugated Ireland*

IT STRIKES ME as strange that our own English people should claim that it was Gurmundus that subjugated Ireland and built the trenches and forts already mentioned, and say nothing of Turgesius whatever. On the other hand the written histories of Ireland and the Irish say that it was Turgesius, and know nothing about Gurmundus. As a result some say that the island was first subjugated by Gurmundus and then a second time by Turgesius. But this is in conflict with the Irish histories. They bear witness that the Irish people were conquered only once before our own time, and that this conquest was by Turgesius. Others suggest that the same man is in question, and that he had two names: the Irish call him Turgesius, while we call him Gurmundus. But again the differing deaths and fates of the two will not allow this to be the case.

The more true and probable account[76] holds that Gurmundus was already ruling in the kingdom of Britain which he had subjugated when Turgesius crossed with a large section of his fleet and a select band of soldiery to this island to conquer it. Turgesius having been the leader of this expedition, and, when the country had been conquered, having remained as the ruler of the kingdom, and as it were, seneschal under Gurmundus, left his name and reputation in lasting remembrance to the Irish people. For they saw and knew face to face him alone, and from him sustained great evils.

113 ¶ *Whence Gurmundus came into Ireland or Britain*

IN THE BRITISH HISTORY[77] we read that Gurmundus came to Ireland from Africa, and thence, having been summoned by the Saxons, besieged Cirencester and took it. Kedericius, then the unworthy king of the Britons, was driven to Wales, and in a short time Gurmundus came to rule the whole kingdom. Whether, then, Gurmundus was an African, or, as is more probable, a Norwegian, either he never was in Ireland, or was there for a short time only and left Turgesius behind him.

114 ¶ *How Gurmundus having been killed in Gaul Turgesius died in Ireland, having been deceived by what appeared to be girls*

WHEN GURMUNDUS, therefore, was killed in Gaul and as a result the yoke of the barbarians was removed from British necks, the Irish people immediately had recourse once again to its accustomed practices in the evil art of deceit – and not without success. Turgesius happened at the time to be very much enamoured of the daughter of Omachlachelinus,[78] the king of Meath. The king hid his hatred in his heart, and granting the girl to Turgesius, promised to send her to him with fifteen beautiful maidens to a certain island in Meath, in the lake of Lochver.[79] Turgesius was delighted and went to the rendezvous on the appointed day with fifteen nobles of his people. They encountered on the island, decked out in girls' clothes to practise their deceit, fifteen young men, shaven of their beards, full of spirit, and especially picked for the job. They carried knives hidden on their persons, and with these they killed Turgesius and his companions in the midst of their embraces.

115 ¶ *About the Norwegians who ruled for some thirty years and how they were driven out of Ireland*

THE RUMOUR immediately went abroad with speedy wings through the whole of the island, and, as is usual, told of the success of the ruse. The Norwegians were cut down everywhere. In short, all of them were through force or guile either killed or driven back in their ships to Norway and the islands from which they had come.

116 ¶ *The wily question of the king of Meath*

THE AFOREMENTIONED KING of Meath had asked guilefully, having already conceived treachery in his mind, of Turgesius how he could wipe out and destroy some birds that had recently come to the kingdom and were a plague to the whole country and the fatherland. When he was told that if they had by any chance nested, all their nests everywhere should be destroyed, the Irish, understanding the remark in terms of the forts of the Norwegians, rose up, to a man, upon the death of Turgesius, to destroy them throughout the whole of Ireland.

The domination of the Norwegians, and the tyranny of Turgesius, lasted for about thirty years in Ireland, and then the Irish people, having shaken off the yoke of slavery, regained its former liberty and again took over the rule of the kingdom.

117 ¶ *The coming of the Ostmen*

NOT VERY LONG AFTERWARDS, however, others came again to the island from Norway and the northern islands. They seemed to be the remnants of the previous people, and knew the country to be very good, either by some deep-seated belief or because they had heard so from their fathers. They came not in a warlike fleet, but in the guise of peace on the pretext of commerce. Immediately they occupied the seaports and eventually with the consent of the chiefs of the land built several cities there.

For the Irish through their vice of innate laziness, of which we have spoken, did not bother to sail the seas or have much truck with commerce; and so with common consent of the whole kingdom they thought it useful indeed that a people who would bring to them the products of other regions and which they did not have themselves, should be admitted into certain parts of the kingdom.

Three brothers led them, Amelavus, Sitaracus, and Yvorus.[80] They first built three cities, Dublin, Waterford, and Limerick. Amelavus ruled Dublin; Sitaracus, Waterford; and Yvorus, Limerick. From them in course of time others went out gradually to build other cities in other places in Ireland.

This people, now called the Ostmen, was at first peaceable and easily managed by the kings of the country. But when they had become very numerous and had surrounded their cities with fine trenches and walls, they began sometimes to rebel fiercely, and to recall the old enmity which they had kept in the depths of their minds. From this and the former coming of the Norwegians the Irish in their anxiety developed the use of the axe. And the deceit from which they had suffered from some, they were only too quick to learn and to employ too frequently on others.

Whence they took the use of axes.

118 ¶ *How many kings reigned in Ireland from the death of Turgesius to Rothericus the last king of Ireland*

FROM THE TIME of king Fedlimidius and the death of Turgesius to the reign of Rothericus of Connacht, who was the last king from among the Irish people, in whose reign and through whom the king of Leinster, Deremetus, the son of Murchardus, was expelled from his kingdom, seventeen kings reigned in Ireland.

119 ¶ *How many kings there were from the first, Herimon, until this last, Rothericus*

THE NUMBER of all the kings that ruled Ireland from the first king of this people, Herimon, to this last, Rothericus, is one hundred and eighty-one. Their names, their deeds, and their times I do not here describe, both because I find few things among them outstanding and worthy of memory, and lest useless prolixity should encumber my account. These kings achieved the kingship of the whole island not through any ceremony of coronation, or rite of anointing, or even right of heredity or order of succession, but only by force and arms. They became kings, each in his own way.

120 ¶ *How the Irish people from the time of its first coming until the time of Turgesius, and from the death of Turgesius until Henry the Second, the king of the English, remained unconquered*

THE IRISH PEOPLE from the first time of its coming and the reign of its first king, Herimon, until the times of Gurmundus and Turgesius, during which peace and tranquillity were for some time upset and destroyed – and again from their deaths to our own times, remained free and unconquered by any attack of foreign peoples, until, invincible king, by you and your courageous daring in these our days it has at length been subjugated.

121 ¶ *The victories of Henry the Second, king of the English*

YOUR VICTORIES vie with the whole round of the world. Our western Alexander, you have stretched your arm from the Pyrenean mountains even to these far western bounds of the northern ocean. As many as are the lands provided by nature in these parts, so many are your victories. If the limits of your expeditions are sought – there will be no more of the world left for you before there will be an end to your activities. A courageous heart may find no lands to conquer. Your victories can never cease. Your triumphs cannot cease, but only that over which you may triumph.

122 ¶ *A brief summary of the same king's various titles and triumphs*

IF THEN YOU BID ME, I shall attempt to describe the manner in which the Irish world has been added to your titles and triumphs; with what great and laudable valour you have penetrated the secrets of the ocean and the hidden things of nature; how you were recalled from a most noble enterprise too soon and inopportunely, too quickly and too criminally, because of an internal conspiracy (although your victory had been won, but the conquest had not yet been put in order); how the petty kings of the West immediately flew to your command as little birds to a light, when they were amazed and dazzled by the light of your coming; how the entrails, as it were, most unnaturally and shamefully having conspired against the belly in a plan that was evil, unjust, and most damaging to the whole Christian world, postponed your eastern victories in Asia and Spain (which you had already decided in your noble mind to add to those of the West and so to extend in a signal way the Faith of Christ); and what mercy and laudable clemency (clemency worthy of being ever remembered and imitated), in the case of a prince and a king most grievously wronged, you, as king and victor, your enemies having everywhere been conquered, showed to the kings and chiefs whom you had subdued.[81]

Truly you are a king and conqueror, ruling your courage by your virtue, and conquering your anger with your temperance, ever keeping in mind that advice of the *Heroides*:[82] 'Subdue your courage and your anger – you have subdued the rest', and turning over in the depths of your mind that noble eulogy of Gaius Caesar: 'The whole world had perished, if mercy had not halted anger', and having often in your hands and always at hand that whole book of Seneca, *On Clemency*, addressed to Nero, and not forgetting that august advice given to Augustus: 'Do as doctors do; when sharp medicines are not successful, they try the contraries.' How gloriously and deliberately you carried out the injunction of that best of orators in your deeds: 'A brave man regards those as enemies who are fighting for victory, but those who have been conquered he treats as men' so that 'bravery may lessen war, and humanity may increase peace.'

If you command me to write the true history of so much and such difficult material, and such as demands capacity far greater than mine, I shall make the attempt. For nothing enjoined by such majesty as yours can seem laborious or troublesome.

'He has won every point who has joined the useful to the sweet.'[83]

NOTES ON THE TEXT

1. *Ecclesiastical History*, I.1.
2. *Polyhistor*, XXV (Salmasius, XXII).
3. Virgil, *Eclogues*, IX. 30.
4. Probably St Modhomhnóg. cf. Whitley Stokes, *Calendar of Oengus the Culdee*, Henry Bradshaw Society, 1905, p. 113.
5. I am indebted to the Rev. Professor Gwynn, S.J., for the following note on this section:

   '[This] elaborate account of the rivers of Ireland . . . throws light on one literary source which Giraldus must have used. It is very different from the lively account of the Welsh rivers to be found in the *Description of Wales*, where Giraldus was writing from personal knowledge. He here distinguishes between the nine oldest rivers of Ireland, and rivers of more recent origin. The list of the oldest rivers is most surprising at first sight . . . This list omits the Shannon, which Giraldus knew to be the main river in Ireland; and also the Barrow, the Suir, the Nore, the Slaney, the Boyne and the Blackwater . . . all of them rivers that ran through lands that had been partially occupied by the Normans before 1185. These are in fact the rivers which Giraldus lists as of more recent origin . . . The explanation of this puzzle is to be found in the oldest extant version of the *Book of the Invasions of Ireland* (*Lebor Gabála*) . . . The oldest version of this legendary history is to be found in the *Book of Leinster* . . . transcribed some twenty years before the coming of the Normans. In the fourth section of this version, which recounts the conquest of Ireland by Parthalón, we are told that there were only three lakes and nine rivers in Ireland when Parthalón came to its shores ... the list of nine rivers is plainly the source from which, in a modified or corrupt form, Giraldus got his list. In *Lebor Gabála* these rivers are enumerated in topographical order: the Liffey, the Lee, the Moy, the Sligo, the Erne, the Finn, the Mourne, the Bush, and the Bann ... Giraldus, or the Latin version that he seems to be following, has altered this order ... the Bush has been replaced by the Saverennus. This last name is no more than a Latin form of the Irish *Sabrann*, an older name for the Lee, probably cognate with the British Severn. A scribe must have added this name as a marginal or interlinear gloss: it has found its way into the text and replaced the Bush ...'

With regard to Giraldus' blunder about the Shannon, one may quote the words of O'Donovan as found in a note on pp. 118 and 119 of *Cambrensis Eversus* in the edition of the Rev. Matthew Kelly (Dublin, Celtic Society), Vol. I, 1848: 'Giraldus thought it possible that a branch of the Shannon discharged itself into the sea at Ballyshannon. The moment you ascend the hill over the Shannon *pot* (*Log na Sionna*), you see the waters all *making for* Ballyshannon. All the streams north of the *Log* flow into Lough Macnean, which discharges itself into the upper and that (in its turn) into the lower Lough Erne, which sends its superabundant waters into the sea through Ballyshannon. Hence some persons have supposed that Shannon (Shanny) was the name of the northwestern chain of lakes; and that Ballyshannon, *Beal-atha Seanaigh*, received its name from being their outlet to the sea. This, however, is not the case; for the oldest name of the river Erne was *Samhair*, and *Beal atha Seanaigh* (Os Vadi Senachi) was merely the name of the ford opposite which the castle Ballyshannon was built.' In any case the Normans had little, if any, precise information about the north-west of Ireland at this stage. One need only look at the maps found in some of the manuscripts of the *Topography of Ireland* to see how true this is.

6. Professor Gwynn has passed on to me a communication from Dr Went which will be helpful here:

'It does seem as if the Pike was an introduced fish; and in this respect I am inclined to accept what Giraldus says. Roderick O'Flaherty, writing in 1684, said that there were no Bream or Pike in counties Galway and Mayo, counties which today provide some of the finest specimens of Pike in Europe. As regards the Perch, there is really no evidence as to whether this is actually a native fish or not, although it is clear that if it is a native fish it has multiplied considerably even in the last two hundred years. There are no native true Roach in Ireland. Roach are only to be found in the Cork Blackwater, where they are known to have been introduced some years ago. The fish called "Roach" in this country is really the Rudd, a very closely related species. The "Gardon" is absent from Ireland. It is presumably the Chub (*Leuciscus cephalus*). The Gudgeon is, however, found in Ireland, but its distribution is somewhat discontinuous. It is certainly now to be found in some of the rivers covered by the counties which Giraldus seems to have visited. The Minnow is also to be found in Ireland, again with a discontinuous distribution. It has been suggested that the Minnow is not a native Irish fish, and certainly within recent years it has spread very much into systems where formerly it was not found. Bullheads are definitely absent. The term *Verones* usually denotes the Minnow, which, however, Giraldus has already mentioned. The Loach is to be found in Ireland, but only one species, whereas Britain has two. As regards the first of the three kinds mentioned as being found only here – this fish is probably one of the Pollan, which are still abundant in Lough Neagh. The Tymal is the Grayling which is not present in Ireland. The Herring-like fish are probably Shad, of which there appeared to be large numbers in the Shannon a hundred years ago or more; and there are still substantial numbers running into Cork Blackwater where they are called "Bony Horsemen". Possibly the third sort of fish mentioned by Giraldus as resembling a Trout may be one of the Char, which, however, do have spots, but of very much different colour and appearance from those of the Trout.' cf. A. Went, 'Giraldus Cambrensis' Notes on Irish Fish', *Irish Naturalists' Journal*, Vol. IX, No. 9.

7. *Variae*, I.24.
8. I have been informed by the Rev. P. G. Kennedy, S.J., that Irish usage tends to confuse the crane with the heron. There are now no native cranes in Ireland but vagrant cranes are occasionally found. Remains of the crane have been discovered in lake-dwellings and other similar sites in the country; and so it is possible that Giraldus may be correct. cf. the account of the 'Catacombs Cave' at Edenvale, Co. Clare, in *Transactions of the Royal Irish Academy*, XXXIII B, p. 55. Higden also in the fourteenth century stated that Ireland abounded in cranes (Ussher and Warren, *Birds of Ireland*, p. 246). The 'wild peacock' is the capercaillie; the 'grutae' are grouse; the 'ratulae' are probably corncrakes (cf. P. G. Kennedy in *Studies*, September 1948).
9. Giraldus has plainly confused the kingfisher with the water ouzel.
10. This opinion was widely prevalent in ancient times. cf. Plato, *Phaedo*, 84e, 85b; Cicero, *Tusculan Disputations*, I.73; Pliny, *Natural History*, X.23, denies it. See also Sir Thomas Browne, *Vulgar Errors*, III.27.
11. Horace, *Epistles*, I.vii, 13.
12. The pheasant is known to be a newcomer to Ireland. It

was introduced in Elizabethan times. The partridge is believed to be indigenous, although there is no native Irish name for this bird.

13. Some modern theory on this matter maintains that the legend of the expulsion of snakes from the country is of Norse, not Irish, origin. It is based on a confusion between the Norse word for 'toad-expeller' (*Pad-rekr*) and the Irish form of Patricius (*Padraig*).

14. Ovid, *Metamorphoses*, XV.375.

15. The account given here of the wonders of Ireland is partly made up of 'yarns', some of them decidedly 'tall', that were told to Giraldus; and partly of tales that are well known from native Irish sources. These latter have a history from long before the time of Giraldus right down to the present day. Mommsen, for example, prints a poem from Paris, *Codices Latini* 11108 in his edition of Nennius (*Monumenta Germaniae Historica, Auctores Antiqui*, XIII), pp. 219 ff., entitled *De Mirabilibus Hiberniae* (The Wonders of Ireland). See also A. Gwynn, *The Writings of Bishop Patrick*, Dublin, 1955. A thirteenth-century report on these same tales is the Norse *Kongs Skuggsjo* (*Speculum Regale*): cf. Kuno Meyer, *The Irish Mirabilia in the Norse Speculum Regale*, *Eriu* IV, 1910, pp. 1–16. Still later accounts are found in the fourteenth-century *Book of Ballymote* and the late medieval miscellany known as *Trinity College, Dublin*, MS H. 3. 17 (cf. Todd, *Irish Version of Nennius*, Dublin, 1848, pp. 192–219). All these versions probably reproduce an oral, if not also a written tradition, going back into the Old Irish period. Similar tales are still being collected in Ireland today by the Folklore Commission. Giraldus claims that he has used no written source for this section: 'in duabus primis nullam prorsus ex scriptis Hibernicis evidentiam ... inveni' (*Topography of Ireland*, Dimock, p. 8). In fact he cites, for example, Solinus and Bede within the first few paragraphs where there are also discernible echoes of

Caesar, *Gallic War*, V.12 ff. and Tacitus, *Agricola*, 10 ff.

16. This section is based on Solinus, XXV (Salmasius XXII). It is discussed in *Cambrensis Eversus* II. 116, where the opinions of Stanihurst, Ware and Boate are quoted against that of Solinus. Lynch typically comments: 'But he was always on the eager watch to fly at anything that might be to the disadvantage of Ireland.'
17. Not extant.
18. Discussed in *Cambrensis Eversus* I. 123. Kelly tells us that: 'The isle of the living was situated three miles from Roscrea, parish of Corbally, in a lake called Loch Cre, which is dried up; but the surrounding bog is still called Monaincha, i.e., bog of the island.' A folklore account mentions Lough Ree as an island on which no female creature, bird or beast, ventured to land in the lifetime of the holy hermit who dwelt there.
19. The Culdees or *Céli Dé* ('Clients of God') seem to have been communities of religious gathered around the reform leaders of the eighth century A.D.; cf. J. F. Kenney, *The Sources for the Early History of Ireland: Ecclesiastical*, New York, 1929, pp. 468–71.
20. See *Cambrensis Eversus* I.141 ff. Lynch contrasts Giraldus' account of St Patrick's Purgatory at Lough Derg (Co. Donegal) unfavourably with that given by Henry of Saltrey who wrote, about the year 1185, a description of a pilgrimage undertaken to Lough Derg by a knight, Owen. Kelly gives much interesting information (loc. cit. pp. 138–55). It is to be noted that this account of Giraldus is the only one according to which the island is divided into sections for good and evil spirits, and in which there is mention of nine pits. cf. Alice Curtayne, *Lough Derg*, Dublin, 1944; Shane Leslie, *Saint Patrick's Purgatory*; and St John D. Seymour, *St Patrick's Purgatory*, Dundalk, 1918; Dublin, 1961.
21. Aran is specified in the later recensions. Other accounts refer this tale to Inis Gluaire off the coast of Erris, Co. Mayo. So, e.g., Lynch in *Cambrensis Eversus*, I.129.

22. The Norse version (referred to in n. 15) places this well on Slieve Bloom in Leinster. The eleventh-century *De Mirabilibus Hiberniae* (see n. 15) has:

    'Cernitur a multis alius fons more probatus,
    Qui facit ut dicunt canos mox esse capillos.'

    Another duly proven well is known
    That quickly greys one's hair.

    The Irish Nennius (195) places the well in Gallorn, Co. Monaghan.
23. The Irish Nennius (197) mentions this well but makes it the source of the Barrow in Slieve Bloom. cf. *Cambrensis Eversus* I.130, 136.
24. Virgil, *Georgics*, II.490.
25. Virgil, *Eclogues*, VIII.63.
26. Lough Neagh. The Irish accounts agree with that of Giraldus, except that they do not mention anything about unnatural crimes as having to do with the origin of the lake. Tighearnach dates the eruption of the lake at A.D. 65–73. cf. Irish Nennius (195). Moore uses the story in 'Let Erin Remember the Days of Old'. The tall, slender, and rounded towers are the famous Round Towers.
27. In *Cambrensis Eversus* I.357, this story is referred to a period four hundred years before the time indicated by Giraldus. The fish was stated to be a whale.
28. Mälstrom, near Lofoten Is.
29. *Polyhistor*, XXV (Salmasius XXII).
30. M reads *Killaraum*, as do other later MSS. Ophela is used by Giraldus to designate the territory of the Uí Faeláin. Uí Faeláin (centre, Naas) is the name of one of the three subdivisions of the kingdom of North Leinster.
31. Geoffrey of Monmouth, VIII.10–12. Stonehenge is in question.
32. 1182 or 1183. The Norse version (cf. n. 15) tells of a

whole race of Irishmen who opposed the teaching of St Patrick and some of whose descendants were punished by being turned into wolves every seven years. Bishop Patrick (cf. n. 15) writes:

> 'Sunt homines quidam Scottorum gentis habentes
> Miram naturam maiorum ab origine ductam'
>
> Certain men of Irish race
> Have from their origin a strange nature

referring to some Irishmen as men-wolves. Other versions of the tale place it in Ossory.

33. 1174.
34. Ovid, *Ars Amatoria*, II.24.
35. XX.16.
36. 1185.
37. Other versions make Cormac son of Cuilennán the planter of an apple-bearing alder-tree.
38. Ferann na gCenél, in the barony of Shelmalier.
39. Virgil, *Georgics*, II.201–2, translation by James Rhoades.
40. Strongbow. The incident is mentioned in the *Conquest of Ireland*, I.17, under A.D. 1170.
41. Raymond-le-Gros, a cousin of Giraldus.
42. In the *Topography of Ireland* (Dimock), p. 8, the author tells us that he had recourse to Irish chronicles: 'in tertia sola . . . aliquam ex eorum chronicis contraxi notitiam'. His brief sketch of Irish history in this third part is undoubtedly based on the oldest version of the *Lebor Gabála*, the 'Book of Invasions'. For these invasions, cf. T. F. O'Rahilly, *Early Irish History and Mythology*, Dublin, 1946.
43. Not identified, although it is said to be in Uí Mic Uais Breagh to the south-west of Tara. cf. Dimock, op. cit., p. 425.
44. Slieve Donard in Co. Down.
45. The inner bay of Dundrum in Co. Down.

46. Lucan, *Pharsalia*, I.511.
47. These quotations are taken from Lucan, *Pharsalia*, I.70–81.
48. i.e. Agnoman, Iarbhainell, Ainninn and Fearghus Leithdheirg.
49. Adapted from Ovid, *Heroides*, XVII.130.
50. Dela, i.e. the Firbolg, possibly the Belgae of the Roman world.
51. In Westmeath. The Hill of Uisneach is meant.
52. i.e. Gann, Genann, Sengann, Rudraige and Sláine or Sláinge.
53. Lucan, *Pharsalia*, I.92.
54. i.e. Gaidel, son of Etheór and Fénius Farsaid.
55. i.e. Girguint and the Basques.
56. Geoffrey of Monmouth, III.12.
57. Henry II had brought this about by his marriage with Eleanor of Guienne.
58. Bull of Alexander III.
59. Lucan, *Pharsalia*, 1.29.
60. Cf. Ovid, *Ars Amatoria*, II.313.
61. See W. H. Grattan Flood, *History of Irish Music*, Dublin, 1927, p. 61 and W. K. Sullivan's introduction to O'Curry's *Manners and Customs of the Ancient Irish*, Vol. I, 490 ff. One played on the timpanum with one's fingers or a plectrum, but on the crowd with a bow.
62. *natione Britannus.*
63. Giraldus is in error here. cf. *Cambrensis Eversus*, III.470 ff.
64. A.D. 1152.
65. This is incorrect. There is a gap of over a hundred years between the traditional date of the death of Patrick (493) and that of Columba (597).
66. Virgil, *Aeneid*, II.390.
67. Irish scholars have tended to reject this whole anecdote as having nothing in its favour from the point of view of historical criticism. Nevertheless Pokorny and Schröder (*Zeitschrift für keltische Philologie*, XVI, 1926, 123,

310–12) have treated it, for the purposes of comparison with surviving relics of Indo-European paganism, as serious historical information. For an account of the inauguration of an Irish chieftain at about the date of Giraldus, cf. O'Donovan, *Kilkenny Archaeological Society Journal*, II (1853), pp. 335–47. Cf. also F. J. Byrne, *Irish Kings and High-Kings*, Batsford, 1973, pp. 17f.

68. This and the two following sections are based on a sermon preached by Giraldus before a synod at Dublin in the Lent of 1186. It was meant as a retort to a sermon preached just before by the Cistercian abbot of Baltinglass, Ailbe Ua Maelmuige, in which he accused (and his accusation was in part proved to the Archbishop of Dublin to be correct) the English and Welsh clergy of bringing with them to Ireland such abuses as existed there. Giraldus, therefore, had a certain interest in painting the Irish clergy in colours as black as possible. In fact, however, some of his accusations are true, and can be explained as resulting from the powerful influence of the provisions of the Brehon Law: cf. A. Gwynn, 'The First Synod of Cashel', *Irish Ecclesiastical Record*, February 1946, pp. 109–22. There had been a strong movement towards reform in the Irish Church throughout the twelfth century. Cf. A. Gwynn, 'The twelfth-century Reform', *History of Irish Catholicism*, ed. P. J. Corish, Dublin, 1968, 2. i.
69. Cf. *Epistle* 125.
70. *Epistle* 14 (not, however, addressed to Eleutherius, but to Heliodorus).
71. *Epistle* 69.
72. Cf. *Epistle* 53. 4.
73. Gratian, D, LXXXIII, I, c. III.
74. *ibid.* C.II, q.VII, c. 55.
75. i.e. the second quarter of the ninth century.
76. Dimock notes: 'The making Gurmund and Tuirgeis contemporaneous is a marvellous exploit in chronology,

even for Giraldus Cambrensis. This African king (Gurmund) . . . must have lived . . . some 300 years . . . before the time of Tuirgeis' (*Topography of Ireland*, 183, n. 3).

77. Geoffrey of Monmouth, XI.8.
78. Murchad Ua Máelechlainn.
79. Now Lough Owel in Westmeath. Herodotus, V.20; Xenophon, *Hellenica*, V.4; Pausanias, IV.4 and Polyaenus, I.20 tell similar stories.
80. i.e. Olaf, Sigtryg and Ivar.
81. All of this is described in Part I of the *Conquest of Ireland*.
82. Ovid, *Heroides*, III.85.
83. Horace, *Art of Poetry*, 343.

READ MORE IN PENGUIN

READ MORE IN PENGUIN 

A CHOICE OF CLASSICS

ANTHOLOGIES AND ANONYMOUS WORKS

The Age of Bede
Alfred the Great
Beowulf
A Celtic Miscellany
The Cloud of Unknowing and Other Works
The Death of King Arthur
The Earliest English Poems
Early Christian Lives
Early Irish Myths and Sagas
Egil's Saga
English Mystery Plays
The Exeter Book of Riddles
Eyrbyggja Saga
Hrafnkel's Saga and Other Stories
The Letters of Abelard and Heloise
Medieval English Lyrics
Medieval English Verse
Njal's Saga
The Orkneyinga Saga
Roman Poets of the Early Empire
The Saga of King Hrolf Kraki
Seven Viking Romances
Sir Gawain and the Green Knight

READ MORE IN PENGUIN

A CHOICE OF CLASSICS

Hesiod/Theognis	**Theogony/Works and Days/Elegies**
Hippocrates	**Hippocratic Writings**
Homer	**The Iliad**
	The Odyssey
Horace	**Complete Odes and Epodes**
Horace/Persius	**Satires and Epistles**
Juvenal	**The Sixteen Satires**
Livy	**The Early History of Rome**
	Rome and Italy
	Rome and the Mediterranean
	The War with Hannibal
Lucretius	**On the Nature of the Universe**
Martial	**Epigrams**
	Martial in English
Ovid	**The Erotic Poems**
	Heroides
	Metamorphoses
	The Poems of Exile
Pausanias	**Guide to Greece (in two volumes)**
Petronius/Seneca	**The Satyricon/The Apocolocyntosis**
Pindar	**The Odes**
Plato	**Early Socratic Dialogues**
	Gorgias
	The Last Days of Socrates (Euthyphro/ The Apology/Crito/Phaedo)
	The Laws
	Phaedrus and Letters VII and VIII
	Philebus
	Protagoras/Meno
	The Republic
	The Symposium
	Theaetetus
	Timaeus/Critias
Plautus	**The Pot of Gold and Other Plays**
	The Rope and Other Plays